Environmental Management

in the

Tropics

An Historical Perspective

Randall Baker
Professor
School of Public and Environmental Affairs
Indiana University
Bloomington, Indiana

CRC Press
Taylor & Francis Group
Boca Raton London New York

CRC Press is an imprint of the
Taylor & Francis Group, an **informa** business

First published 1993 by Lewis Publishers, Inc.

Published 2019 by CRC Press
Taylor & Francis Group
6000 Broken Sound Parkway NW, Suite 300
Boca Raton, FL 33487-2742

© 1993 by Taylor & Francis Group, LLC
CRC Press is an imprint of Taylor & Francis Group, an Informa business

First issued in paperback 2019

No claim to original U.S. Government works

ISBN-13: 978-0-367-45010-6 (pbk)
ISBN-13: 978-0-87371-661-1 (hbk)

Visit the Taylor & Francis Web site at
http://www.taylorandfrancis.com

and the CRC Press Web site at
http://www.crcpress.com

Library of Congress Cataloging-in-Publication Data

Baker, Randall.
    Environmental management in the tropics: an environmental perspective /
    Randall Baker.
        p.   cm.
    Includes bibliographical references (p.   ) and index.
    ISBN 0-87371-661-2
    1. Environmental policy—Tropics—History.   2. Land use—
Environmental aspects—Tropics—History.   3. Human ecology—
Tropics—History.   I. Title.
HC695.Z9E52   1992
333.7′0913—dc20   92-25722
                                                                            CIP

Preface

When the Chinese say "May you live in interesting times", they are doing you no favors, as this is intended to be a curse. The Chinese Empire, which had survived in one form or another for almost five millennia, put a high premium on stability, to the point where the calligraphy next to the Heavenly Throne read "change nothing".

This, however, is a book about change — rapid and externally induced — that transformed an entire region of the globe more or less coincidental with the tropics and subtropics. It is that process of change that really lies at the heart of this study. In the first place it is difficult to discuss change unless you know what was there to start with. In the context of this book, that involves the interaction of what are now rather patronizingly called "traditional cultures" with the available natural resources their technology enables them to exploit. To discern this interrelationship is difficult since much of the evidence has gone or has been seriously distorted, because much of the written record was provided by biased outsiders, and because the traditional land-use systems were subject to the sudden onslaught of European religion, science, culture, and perception.

The unequal coming together of these two worlds, supposedly "traditional" and "modern", was to totally and irrevocably change the relationship of people and nature, and indeed people and people, across the tropical realm. Now we accept without question that nearly all the poor countries are tropical, but why should this be so? We accept the idea that two people doing the same agricultural task in the tropics and the temperate world command significantly different reward levels for their effort. Why is this? Is it environmentally determined?

The tropics, environmentally and economically, are in trouble, and the only satisfactory way to understand why, and consequently what may be done about this situation, is to study the historical evolution of the changing nature of environmental management in this part of the world. At the same time we have to recognize that the tropics has no more homogeneity than does the "Third World". However, almost all tropical countries shared a common historical experience of being suddenly separated from the indigenous momentum of their own history and being thrown into someone else's back yard. This happened at an absolutely critical time in the history of the colonial powers as they embarked on an unprecedented rapid transformation to urban, commercial, and industrial societies based increasingly on experimental science.

The central thesis of this text is that history and a melding of artificial disciplinary divisions are essential for gaining any real understanding of the environmental tragedy and continuing dilemma of the tropics. This is not an exposé of the guilty or a search for blame. It is a journey toward a better understanding of some truly terrible realities.

To achieve this we have to marry some natural science, some anthropology, a lot of history, and indeed anything else that matters. The aim is realism, understanding, and a little humility in the face of some gigantic myths and even more gigantic problems. At the end of the book perhaps we shall know better how to face the future once we know where the present came from.

Randall Baker
Bloomington, Indiana

The Author

Randall Baker was born in Wales in 1944. He studied geology, economics, and geography at the University of Wales in 1965. Pursuing a long-established interest, he went on a Rockefeller Grant in that year to Makerere University College, Kampala, Uganda. There he studied natural sciences, African history, and agricultural economics. His doctoral thesis as a Goldsmiths' Scholar was on the environmental effects of attempting to commercialize predominantly nonmonetary economies, concentrating on cattle-keeping communities. This resulted in the award of the doctorate in 1968 from the University of East Africa.

He remained at the University of East Africa for another 2 years.

In 1970 he went to the new University of East Anglia in Norwich, England to join the Environmental Science faculty. However, in 1973 with several colleagues he was a co-founder of the School of Development Studies, which combined natural and social sciences within one faculty around a policy focus. At the same time he held a position at Cambridge University. In 1975 (at the age of 31) he became dean of the school, and then was at UNESCO (Paris) working on an arid-lands policy and UNESCO's position paper at the UN Conference on Desertification in Nairobi. In 1982 he went as the European Community adviser to Fiji on economic planning ending up as Special Advisor to the Minister of Primary Industry.

In 1984 (at the age of 40) he decided to accept a chair at Indiana University's multi-disciplinary School of Public and Environmental Affairs. Here he concentrates on bringing the comparative and international dimension into professional education in both public affairs and environmental management. Here he introduced the theme of environmental management in the tropics to provide new insight on the complexities of society and environment.

He has worked on all the continents and has published books in six languages, including two recent works on government in small countries and comparative public management.

His current work includes a joint Russian-American book on the Aral Sea crisis and advisory efforts on developing environmental legislation and structures in Slovenia and Bulgaria. In 1992 he was awarded a Fulbright scholarship to help develop public administration and environmental management in Bulgaria. This is connected with the growth of the New Bulgarian University. This work, in turn, builds upon his efforts in developing the private University of Bolivia in Cochabamba.

Contents

Acknowledgments

The origins of this study lie 25 years ago when I was a student at the University of East Africa at its Makerere campus in Kampala, Uganda. This was an inspirational academic enviroment, and I owe a great debt of gratitude to my professor, Bryan Langlands. This book results directly from his inspiration. At Indiana I owe a special debt to Lynton Keith Caldwell, whose comments on the first draft helped resolve some basic issues in my mind. Many generations of students provided me with constant comment and reaction enabling refinement of the ideas. Throughout the preparation of the manuscript I have received steady assistance from two graduate assistants, Tom Bayer and Frank Novak, whose care and attention have been very significant. My special thanks, however, are reserved for my wife Susan Baker who, by one of those divine accidents, is a very skilled and ruthless copy editor. Publishers and authors were universally forthcoming with their permission freely to reproduce their work.

A Perspective on this Work

This is an important book. I hope that it will be read widely — especially by people engaged in so-called foreign aid, international technical assistance, and international business. *Environmental Management in the Tropics* concerns not only the greater number of people on earth, it also probably includes roughly half of the inhabitable land surface of the globe. As Randall Baker rightly observes, the tropics is not a region to be defined in precise or general categories. Even its deserts and rain forests are distinctive and different, no less its peoples and their cultures. The term "tropics" is essentially a modern West European concept which, like "development", "progress", and "Third World", is an unwarranted extrapolation of Eurocentric assumptions to principles or perceptions believed to have universal applicability.

This book is an effective antidote to a planetary provincialism that, through good intentions as well as through rationalized exploitation, has brought destruction and degradation to peoples and societies around the latitudes of the earth called "tropical". There have been benefits from the extension of "modern" civilization to the tropics — even to some peoples in the tropics. But the "goods" largely redounded to the Western countries or colonial powers while the "bads" tended to accrue in the tropics. However, the continuation of this 500-year-old historic relationship has become untenable.

The title of this book might with equal, perhaps greater, propriety have been *Environmental Mismanagement in the Tropics*. Mismanagement is what much of the text of the book documents. But the reader should bear in mind that environmental mismanagement was not, nor is, confined to the tropics. The temperate regions have generally suffered less from technoeconomic and managerial folly, in part because climatological and geological conditions made those regions less vulnerable to rapid deterioration. Also, it seems that the societies that invented modern technologies were better equipped to adapt to them. The industrial mode of production originated and was elaborated in their midst and evolved over time sufficient for adjustment to their imperatives.

The book provides a pertinent summary of what must be understood if "sustainable development" is to be more than a slogan. The modern world is not responsible for all of the ills that have befallen the tropics. Tropical societies often made themselves vulnerable to European intervention. One should not forget that the advanced civilizations of the Mayas and Khmers collapsed without apparent external causes. In Africa, slave-raiding tribes delivered their fellow blacks to the dock to be sold to non-European, non-Christian Arab slave-traders as well as to Europeans. A microcosm of the plight of the tropics may be Easter Island, where the inhabitants, lacking foresight, first destroyed the viability of their environment and then, turning on one another, destroyed themselves when the Europeans arrived to help finish the job.

This book does not provide a blueprint for sustainable environmental management in the tropics. It does provide a basis for an urgent and essential ecologically rational reorientation of development for which learning (and unlearning) will be necessary. The historical perspective that the book provides stands out as being exceptional among a large number of development publications — chiefly by "development economists" — which are without historical or ecological understanding. There is no longer a valid excuse for failure to understand why and how the world arrived at its present predicament. Baker does not apportion blame or demand retribution for the sins of colonial exploitation. His focus is on the present and future. His viewpoint recalls that expressed by Abraham Joshua Heshel, who wrote: "In a free society some are guilty, all are responsible." In the world of the 21st century, the survival of free society and perhaps even of humanity will require a universalizing of a sense of responsibility for human/environmental relationships. This book gives rationale and impetus to the imperative.

Lynton Keith Caldwell
Arthur F. Bentley Professor Emeritus of Political Science
Professor Emeritus of Public and Environmental Affairs
Indiana University

CHAPTER 1

Some Common Themes

Strictly speaking, the tropics is that part of the earth's surface bounded by the Tropic of Cancer (23° 27' north) and the Tropic of Capricorn (23° 27' south). This zone of the globe is delimited by the extremities of the movement of the sun relative to the earth through the seasons. This, in turn, results from the fact that the earth does not revolve around a vertical axis, but a tilted one, in its journey around the sun each year (Figure 1.1). The implications of this are considered more fully in the coverage of the tropical climate in Chapter 2; in brief, it means that the tropical area receives a high intensity of solar energy — the sun's rays strike the globe at a high angle — resulting in the characteristic warmth of the area when not tempered by cold ocean currents or mountains. This year-round warmth is the defining feature of the tropical environment, though in terms of its usefulness it is sharply modified by the availability and seasonality of moisture.

For the purposes of this book, however, it is expedient to be rather more liberal in our definition of the tropics since it is rare indeed for nature to observe a straight line on the surface of the land or the oceans. We shall include considerable lowland areas, of what might be called the subtropics, where the major vegetation zones blend imperceptibly north and south into more temperate, distinctive plant types. This approach is more realistic in terms of the land use interests of this work. Perhaps the best indicator would be the extent of tropical **climax** vegetation types — those that would occur naturally in low-lying areas without the modifying influences of human intervention — as shown in Figure 1.2. There are, as we shall see, many gray areas along the margins of the tropics, and indeed these are very important since it is in these areas that some of the most important modifications of climate and vegetation

1

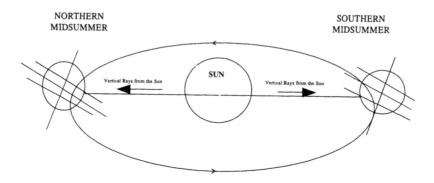

Figure 1.1. The origins of the tropic lines and seasons.

are occurring. Such a boundary situation is exemplified in our study of desertification (Chapter 15), which is often a classic interzonal, as well as intrazonal, problem. Unfortunately, from this definitional point of view, people have been around for a very long time and, since the Neolithic period (circa 10,000 years before the present), they have been active in clearing, burning, cultivating, and grazing vast areas. So, it is not always easy to reconstruct the climax vegetation from the remnants that remain on the ground. Nevertheless the climax concept remains the best guide as to what would happen if the interplay of natural forces was left to itself in major ecological regions.

Throughout history, the tropical regions have been the subject of many broad misconceptions by the people of the temperate world who were, eventually, to become political masters of the tropics during the colonial period, roughly between the 16th century and the 1960s. Even in the times of ancient Greece, there was a general belief that people venturing into the "torrid zone" would eventually become deranged. In part this was due to the fact that anyone venturing south from Europe had a number of inhospitable routes to follow. One was down the west coast of Africa along the Moroccan and Western Saharan coast, which is about as barren a place as one could find anywhere. The route south faced the daunting prospect of the Sahara itself, which swallowed up at least one entire Roman expedition without trace. On the east lay the Red Sea, with its arid coastlines on both the African and Arabian sides.

Later, when Europeans reached the humid tropics, they quickly fell victim to the debilitating and killing diseases that plagued them in those areas. There were, of course, devastating diseases in Europe, but Europeans had little if any resistance to the tropical plagues and they died of them in great numbers. West Africa came to be known in the 18th and 19th centuries as the "white man's grave". We tend to forget now that droves of indigenous people and their animals in the tropics died from imported "European" diseases, such as smallpox and rinderpest, and continue to do so in the Indian areas of the Amazon today. The people of the tropics would have been equally justified in thinking of Europe as a place of terrible pestilence and generally inhospitable to human

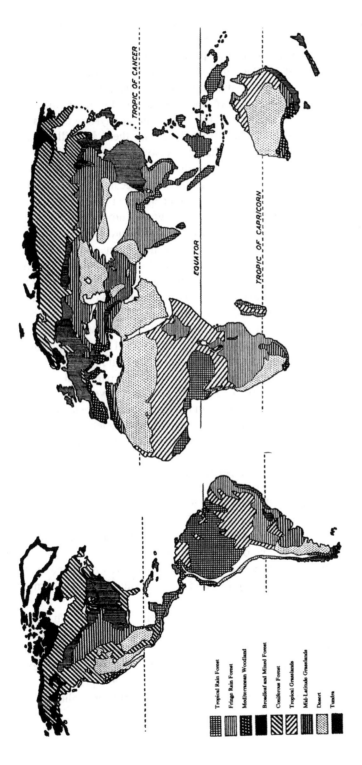

Figure 1.2. Major world vegetation types.

life. Medieval Europe pictured the tropics as a place of rumored fabulous wealth (the myth continued into Rider Haggard's *"King Solomon's Mines"*) and of fantastic creatures such as people whose heads were located in the middle of their chests! In fact, it is worth remembering throughout this book that vast areas of the tropics remained essentially unexplored by Europeans (or "undiscovered", as we would have said until recently) well into the latter part of the 19th century. It was only in the 1860s that the mystery of the source of the Nile, for instance, was finally resolved, and there are still parts of the Amazon harboring essentially unchanged lifestyles and countless thousands of undocumented species of plants, animals, and insects.

All this discussion of perception is important because of the unequal relationship that came to exist between the temperate and tropical regions during the colonial period and the tremendous flow of European ideas of science, religion, administration, and "civilization" southward (and even northward from Australia to New Guinea and from the white colonies of South Africa into the tribal areas). This legacy of "perception" was as least as important in understanding what happened to the tropics as anything related to "objective scientific observation and experiment". Hence, readers will find that this book on the tropical environment has a rather large dose of history and myth interwoven into the text. It is almost impossible to understand what occurred without reference to past perceptions and, as we shall see, much of this myth persists, especially in the use of words such as "traditional".

By the early years of the 20th century, a substantial part of this folklore was being codified into "scientific theory". The persistent poverty and low productivity (defined by output per person and per acre) was attributed to the debilitating effects of the climate and its associated disease patterns. This was the era of **environmental determinism**, which was associated with such geographers as Ellsworth Huntington and Ellen Semple, who both proposed a "natural explanation" for the broad and persistent "backwardness" of this region of the globe. According to Huntington and Semple, a harsh climate and disease sapped the capacity of the people to work, particularly at times of seasonal stress, thereby reducing the amount of food they could produce. This low productivity limited the opportunities for differentiation to occur in society, which is often taken as one of the bases of the emergence of civilization (the priests, builders, mathematicians, etc.). Low productivity restricted the formation of capital and the ability to build a better future. This in turn caused poverty to persist (see Text Box 1.1).

The ideas expressed in these circular arguments became known as "vicious cycles of poverty", and were said to "explain" the origin and persistence of primitive forms of land use and the low level of "social development", as well as the "nasty, brutish, and short" lifestyles. Superficially, this facile explanation seems to satisfy many lines of inquiry, and it also provides a formula for change and improvement — the transfer of temperate values as part of

Text Box 1.1
An Environmental Determinist View of the Tropics

Ellsworth Huntington tried to provide a "scientific" basis for the persistent backwardness, as he saw it, of the tropics. The critical factor in his view was climate. In his day his writings were persuasive, but now they seem to border on the incredible. Here are two extracts from his *Climate and the Evolution of Civilization* presented to the Yale chapter of Sigma Xi in 1916 and published by Yale University Press in 1923.[1]

pp. 149-50: "In tropical regions the energetic types unfortunately kill themselves by overexertion. Activity accelerates the processes of metabolism and generates toxic poisons. In the right kind of climate these are eliminated during periods of rest. In a warm climate, however, the high temperature appears to cause excessive chemical activity of the protoplasm just as does exercise. . . . The exact mechanism of the process has not been determined, but some poisoning of the system and consequent elimination of unduly active types appears to be one of the reasons why the negro has acquired a comparatively indolent character." [Unbelievably, the author never mentions the diseases introduced by Europeans or the 300 years of slavery!]

pp. 182-83: "In these later days since historical records have been kept, the contribution of tropical countries to civilization has been as meager as appears to have been the case in the remote past. So far as we are aware, no truly great man except Mohammed has arisen within twenty-five degrees of the equator. . . . In modern times it is even harder than in the past to find great men who lived in tropical or even warm countries. Diaz in the high, cool, but nevertheless tropical plateau of Mexico may be cited as an example, but a hundred years from now, only the special student will have heard of him, while men like Lincoln, Pasteur, and Humboldt will still be admired by thousands, yea, millions of people. [Diaz will, presumably, be forgotten]. . . . Tropical countries may indeed have made no appreciable contribution to civilization at any time, but no later than two or three thousand years ago many countries where climatic energy is now comparatively low [by this he means the tropics] were the seats of the highest civilization. How was this possible? The answer is that a great mass of evidence [?] seems explicable only on the hypothesis of pulsatory changes of climate during *historic times*. [My emphasis.] This evidence has been discussed so fully in other publications that it seems unnecessary to repeat it here. [He then cites a series of his own articles as evidence.] It will be enough to state the main conclusions with almost no details of proof."

a civilizing mission. For others the absence of Christianity provided an explanation, with a similar swift road to recovery. And so there has been a strong unidirectional view of the tropics: it is a place requiring not so much understanding as benign advice and a hefty dose of paternalism from the temperate zone. We shall explore this attitude in more detail when we look at the coming of the Europeans in Chapter 8. But for now it is important to understand that prejudice and myth have been powerful influences in shaping the unstable way that people have come to relate to the land in the tropics over the last 300 or so years.

Essentially, the mind-set of Europeans in relation to the tropics has, therefore, been a *negative one*. The focus has always been on the constraining factors, both in the natural environment and the organization and beliefs of indigenous societies. This is well reflected in what was the standard work on the tropics for a generation: Pierre Gourou's *The Tropical World*.[2] Readers are invited to browse through this book, which only recently went out of print, and consider the chapter headings and the generally pessimistic tone. It is, perhaps, one of the last survivors of the "environmental determinism" school, though it is unlikely that its author would agree with this definition. As late as 1976, Andrew Kamarck was writing on behalf of the World Bank that "the population [is] less vigorous through disease, and possibly, through the direct impact of temperature and humidity."[3]

It is important, at this point, to outline some of the underlying questions and themes of this book. They are not dealt with systematically in chapters, but permeate the rationale for the whole exercise. It is also worth pointing out that this book possibly may not answer all these questions, but it does have the aim of raising them and confirming their legitimacy. Often, as Francis Bacon observed 400 years ago, it is as important to know what questions to ask as it is to provide answers to them.

Why is there a Broad Correlation between the Countries in the Tropics and the Incidence of Poverty and Poor States?

We have already said that the tropics is a geographically bounded concept arising from a purely natural phenomenon (the movement of the earth around the sun on a tilted axis). However, if one were to take a good physical atlas, and at the same time take the current edition of the *World Bank Atlas*,[4] something remarkable would be evident right away. The vast majority of the less developed (poor) countries on the globe lie in this region, or have extensions into this region (Figure 1.3). This is especially true if we are a little more liberal in our definition of the tropics to include areas 5 or 10° north of the tropic of Cancer and south of the tropic of Capricorn (Figure 1.4). There are exceptions, of course. Singapore is a tropical country, and China (a rather large exception) is not, and we shall have to deal with this. However, the poorest countries of Latin America, Africa, and Asia seem to bear out the general observation. It is interesting to note, also, that in the case of South America, the

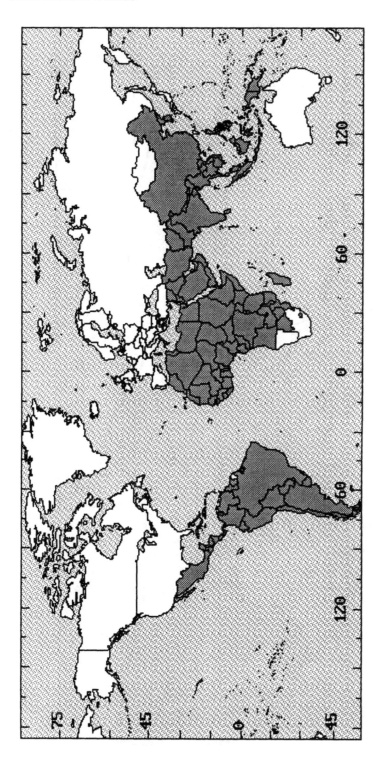

Figure 1.3. Developing countries. (Map courtesy of PC Globe, Inc., Tempe, Arizona, U.S.A. Copyright 1991.)

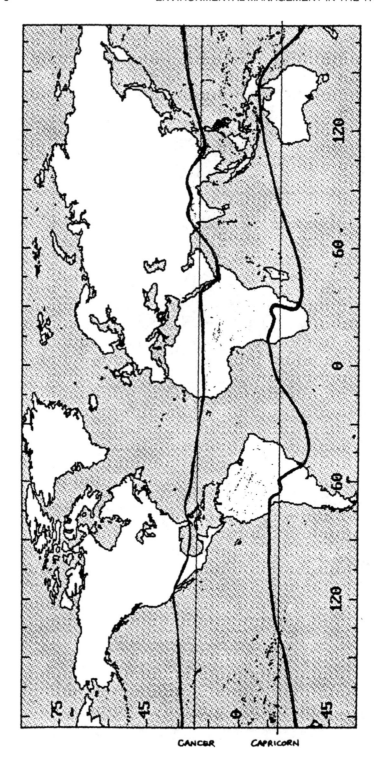

Figure 1.4. A more "liberal" view of the "economic tropics". (Map courtesy of PC Globe, Inc., Tempe, Arizona, U.S.A. Copyright 1991.)

coincidence used to be even greater. Before the Second World War, Argentina was the 10th-most developed country in the world. Both Argentina and Uruguay have *regressed* so that now they show up in the World Bank's "developing nations" category. Here, in the so-called "developing nations", we find the lowest life expectancy levels, highest infant mortality, lowest levels of literacy, and highest incidence of malnutrition, as well as periodic famines. The provision of basic human needs is lowest for the broad spread of the population in this intertropical zone.

This broad statistical and geographical correlation raises some fairly broad causal questions right away. Chief among these is: "Is it not the case, therefore, that the cause of poverty and degradation *is* the physical environment of the tropics?" This echoes the Gourou/Kamarck viewpoint outlined earlier. Indeed this is almost like Marxism in its sweeping and comprehensive explanatory strength. This question really *must* be examined since it seems fundamental to an explanation of cause and effect, and will be basic to the determination of "solutions".

In much of the economic writing on development, poverty — which as we shall see is a major causal factor in, and not simply a result of, environmental destruction — simply *is*. In most of the texts, there is no examination of why it is where it is. It simply *exists*. A whole series of essentially "technical" remedies is then proposed (capital loans, technical assistance, training, etc.). On the other hand, economists and others adhering to the "dependency" school attribute the poverty of the poor nations to the fact that they were entrapped in the unequal colonial system and its economic and political successor institutions, so that they serve the "center" from the "periphery". This, again, does not really tell us why the "center" happens to be in the temperate parts of the world and the "periphery" in the tropics. It is true that there are poor people in rich countries and rich people in poor countries, but we have a broad, global coincidence that is hard to ignore.

A corollary is the idea that the success of the tropical world has largely been measured by the standard parameters of Western "success". These parameters are based on the monetary exchange economy as revealed in the gross national product (GNP), which is a country's total value (in money terms) of the flow of goods and services over a year. This assumes, more or less, that everything is done for money. There is also the GNP per capita, by which the first figure (usually expressed in U.S. dollars) is divided by the total number of people in the country. GNP per capita gives us a broad measure of productivity, again within the money economy, and allows us to construct a world league table of productivity. But GNP per capita takes little to no account of the nonmonetary economy or the way that people are using resources (are they gobbling up capital or drawing down interest?). These issues will be discussed later in an examination of what constitutes development in environmental terms (Chapter 10) and even of whether the globe could sustain the "development" of the tropics in the manner conventionally defined as modernization (i.e., becoming more like the West).

It is important to recognize that we have grown uncritically accustomed to the idea that poor countries are generally hot places. For some this argument has racial overtones. In case we think that this is an issue restricted to extremist groups, it is worth remembering that this was conventional wisdom some years ago. Consider, for instance, this opening paragraph from the *Atlas of Empire* published by Alfred Knopf in 1937:

> Europe is primarily, of course, a group of States holding colonial possessions in other continents; its peoples belonging to the superior white race which has been entrusted with the mission of civilization in other parts of the world (J. F. Horrabin 1937, 3).

Even today, the extreme poverty of Africa is, for some people, confirmation of the inability of black people to develop themselves, except in those countries that have held onto the European "civilizing" model and its values (e.g., Côte d'Ivoire, Kenya, Malawi). This does not take into account the majority of African countries that have failed largely through the adoption of another European model: socialism and central planning. Latin America, according to the racial paradigm, is "better" than Africa because it had a stronger European settler influence. Asians, especially Japanese and Koreans who are more temperately located, are essentially well placed because they are light-skinned, after all. This appears nonsense to many of us, but it no doubt lies behind the thinking of a lot of people even though they would never admit it, sometimes not even to themselves. We really have to expose the lie of this type of "explanation" if there is to be a reasonable and rational assessment of the future of the tropics, its people, and its environmental resource base. So the question "Why are these places poor?" is absolutely central to this book, and this is why, even though it is supposed to be a book on environmental management, this book contains a strong dose of the "soft sciences" so often regarded with disdain or contempt by the "hard" science folk. We may even ponder the question: "Why is this question almost never asked?" Given a geographical coincidence of this magnitude, one might have expected the grants and research to have flowed like Father Nile. Not so.

Why Environmental Management in the Tropics? Ecology is Ecology and Environment is Environment Wherever You Find It

It is certainly true that one of the missions of the book is to apply sound ecological and environmental management principles, which are generic in nature. However, the management concept is an interplay of socioeconomic and natural variables with a cultural context. In the tropics, the form and function of some of these variables is not always immediately obvious to the foreigners who tend to write the books (including this one) and reports, and who advise on legislation and land-use practices. There is no attempt here to define a unique set of processes, simply to be clear about how the basic systems of natural resource use occur in the tropics. The breakdown of "traditional" grazing systems and areas in Africa, for instance, is not a result of a unique *process* of ecology; it is the result of the introduction of temperate, market-oriented management concepts and methods into nomadic, and only partially monetized, economies. It is also not a result of the people being *irrational* or *perverse,* as economists used to say in the 1960s. If anything, the economists were being irrational and perverse in assuming that the people of northeast Uganda marched to the same drum as the people of Wyoming.

The relevance of the subject that this book tackles derives in large part from the fact that the people of the tropics live closer to nature than do the people of the temperate zones, who have increasingly insulated themselves, with capital and energy inputs, from the immediacy of nature for the last 200 years. Indeed farmers, hunters and gatherers, pastoralists, and fishermen constitute the vast majority of the people in the tropics. Thus, what happens to the rural environment in this part of the world has a sudden and pervasive impact on millions of people, as has been seen in Brazil and Ethiopia, as well as in many other less-publicized cases.

Frequently the natural resource base of the tropics requires *different* management techniques from its counterparts in the temperate zone. We shall see this in forest areas where the great weight of vegetation belies the idea that it indicates a great store of natural soil fertility. Generally the fertility is locked in a short cycle of reuse based on the vegetation itself and a few centimeters of what might be called "soil". This Western misapprehension has plagued the "development" of forest areas and led to wholesale degradation, which continues today in the Amazon, where forest is being replaced by miserable grain production and livestock raising (Chapter 14).

Again, through deforestation, soil erosion, and desertification, large areas of the tropics are being lost, and their replacement time — if it is ever even allowed — runs into centuries. This loss, bearing in mind the immediate reliance of large sections of the population on the land for subsistence, is critical indeed and needs urgent attention. If we are to avoid Malthusian doomsday scenes such as we see on television from Ethiopia, we have to institute sound environmental management practices that are appropriate to the realities of the socioeconomic circumstances of these countries.

The Tropical Environment is Subject to Intense Pressure: Can It Sustain This?

We shall discuss Thomas Malthus and his hypothesis that there is an essential contradiction between the fixed nature of natural resources and the seemingly unlimited potential for the growth of human populations in Chapter 7. The poor, tropical countries are those very places where we find the most rapid rates of population growth often giving rise to the destruction of the environment. In some cases it is true, as in Kenya or Central America, that the destruction might also be a result of unequal access of the population to land.

However, a simple projection of gross numbers shows the present situation to be unsustainable, whatever the local social injustices. Some demographers argue that the only way to lower the rate of population growth is through increased economic security, literacy, and changes in the role of women. This might seem difficult to achieve in a place like Africa, where the basic indicators of growth and well-being have been in absolute decline for a couple of decades, or in Mexico, where real wages have been declining steadily for a decade.

At the same time, for many people, children offer the only real form of insurance from want in old age, and so we are confronted with a real paradox. Nevertheless, the "population problem" has to be tackled, however it is defined, if the destruction of resources is not to accelerate. This may involve moving people from the land to other sectors of the economy or making agriculture more productive (the "green revolution"), along with all the environmental problems associated with global increases in the use of pesticides, fertilizers, and fossil energy. It is true that as things stand now, the world could feed itself. Indeed the United States actually pays farmers to keep land *out* of production. The problem is due to the fact that a large, and growing, proportion of the world population lacks the economic base to buy this surplus production, to gain access to productive land, or to grow its own food without damaging the environment in serious ways.

Do We Really Understand the Meaning of the Past in the Tropics?

When the Europeans first came to the tropics, they rarely considered indigenous methods of cultivation or "ethnoscience" worthy of study, and so little is known about "traditional" management and coping mechanisms.[5] As we have observed, most of our ideas about the tropics have been derived second-hand from biased or prejudiced observers who came from radically different environments. The main concern of those who came was to *change,* rather than study, the people and systems they encountered. However, we shall see that those indigenous people had often constructed elaborate systems of natural adjustment and land use (Chapters 6 and 7). These often collapsed under the pressure of rapid, induced, and totally alien change.

Here we have another paradox, because by the time influential writers such as the Scottish missionary David Livingstone came to write about "his people", they had already undergone several centuries of externally induced trauma, which had entirely dislocated their societies through the slave trade. What he was observing was therefore a society in ruins. But often what was seen was taken as the natural state of affairs, and so the vocabulary of the tropical world — "barbarism, heathenism, backwardness, simpleness, primitiveness", etc. — came to be the orthodoxy of the day and a profound explanatory system in itself. Still, if one asks the students of today for the key words that they would use to define the tropics, these are the terms many mention. Even if the tone is paternalistic rather than prejudicial, it really does not matter; the damage is done.

The fact is that the Incas, Mayas, Aztecs, Javanese, the people of Engaruka (Tanzania), and many others, including the Egyptians, constructed the archetypical early civilizations, and they did it in the tropics. This rather takes the wind out of the sails of the environmental determinists. However, they may then argue that these societies did not endure (though the Egyptians kept going longer than Europe has shown a capacity to sustain itself). The fact is that many of these societies collapsed under the weight of external forces. Thus the wonderful hydraulic civilization of the Nabateans of the Near East was done in not by inherent weakness, but by the Romans. The civilizing mission of Spain put an end to the Inca system, and so on. At this point in history, it is not clear how much indigenous wisdom, or ethnoscience, is left to be retrieved in the hybridized societies. But it is at least a worthwhile exercise to try to understand how it was that these societies persisted in their precapital circumstances for so long (though some of them created massive capital infrastructures, such as the terraces of the Philippines or the irrigation networks of Egypt, through centuries of human endeavor). At least we should be prepared to learn.

How Homogenous is the Tropics?

We are as lax about differentiating the tropics as we are about subdividing the so-called developing countries. In truth, the tropics, poor though it generally is, contains a wide diversity of natural ecosystems, and we should recognize this. Again, when most people are asked to set forth the defining terms of the "tropics", they usually think of the "jungle" scenario (hot, humid, forests, lions, . . .). In fact the humid tropics (what Gourou called the "hot, wet tropics") makes up less than half of what constitutes the tropics. The remainder contains everything from savannas to the vast arid and semiarid extent of the Sahara and Sahel. For most of the people living in the tropics, the constraining factor is the seasonality and availability of moisture.

Is the Tropical Environment Changing?

We have already observed that the rural environment of the tropics is subject to intense pressure from population growth leading to degradation, deforestation, and desertification. However, apart from these anthropogenic changes, questions arise from current hypotheses about global environmental change. There is much speculation about global warming and what that would do to the broadly latitudinal agricultural belts of the temperate zone. There is less discussion about what global warming would do to the monsoonal regions, home to tens of millions of people, or to those marginal environments such as the Sahel, which could be rendered uninhabitable by relatively minor changes in the total amount or the seasonality of precipitation (Chapter 2). This clearly has major implications for the study in hand, and we shall try to survey the evidence, such as it is, about possible trends that will have to be accommodated in the tropical environment. This is so important because the availability of capital to reduce the impact of global change by irrigation, migration, and crop genetics is much more limited in poor countries. People in poor countries are far less "insulated" from the immediacy of the environment than those in the temperate zones. Productivity is already low and the resources for capital investment are severely constrained.

What Does Development Mean?

Clearly, in the context of this book, development means finding a sustainable solution to the management of natural resources in the tropics. While this is a sort of guideline, its implementation is very problematic. One possibility would be to limit population growth through draconian measures. This is unrealistic and unlikely to occur until other conditions about family security are satisfied. We could, on the other hand, sustain the population by giving away food surpluses, for a while at least. But, as was seen in India under the PL480 U.S. food assistance program, this tended to lead locally to a sweeping of the basic problem under the carpet, rather than producing any real solution. In fact it provided the means whereby the incipient agricultural/food crisis could be avoided in the policy arena. As the UN Commission on the Environment and Development stated, real sustainability means securing the availability of natural resources for future generations, rather than asset stripping them for "growth today". It is taken as a sine qua non in this study that induced change, or "development", embodies this concept of resource perpetuation. This may involve a substantial rethinking of the conventional "modernization" models of stages of economic growth, according to which there is a unilinear pathway of "development" by which people essentially emulate what happened in the West. Meanwhile, in the West, society is seriously evaluating the terrible costs of the urban/industrial growth and development model in terms of the breakdown of traditional social bonds, the growth of crime, and the wholesale destruction of the natural environment.

And so the scope of our study is interdisciplinary, and is therefore likely to annoy many specialists who feel that "sorghum has got short shrift", or who ask, "what about methane digesters?" The fact is that this book is intended to be a broad interdisciplinary review, and no apology is made for introducing religion, international economics, politics, and a variety of other factors into a book bearing the title "environmental management". Furthermore, the focus of the book is essentially rural. Perfectly horrible urban environmental problems exist, too, and possibly the approach taken here might inspire someone else to write about those. The fact is that the overwhelming majority of people in the tropics are *rural*, and someone has to feed the town dwellers, who all too often live in a parasitic relationship with their country cousins because of a pervasive predominance of the modern, urban sector in orthodox development models over the last few decades.

REFERENCES

1. Huntington, E. "Climate and the Evolution of Civilization," in *The Evolution of the Earth and its Inhabitants*, J. Barrell, C. Schuchert, L. L. Woodruff, R. S. Lull, and E. Huntington, Eds. (New Haven: Yale University Press, 1923), pp. 147-93.

2. Gourou, P. *The Tropical World: Its Social and Economic Conditions, and Its Future Status* (New York: John Wiley & Sons, 1966).

3. Kamarck, A. M. *The Tropics and Economic Development: A Provocative Inquiry into the Poverty of Nations* (Baltimore, MD: Johns Hopkins for the World Bank, 1976).

4. World Bank. *The World Bank Atlas* (Washington, D.C.: The World Bank.)

5. Richards, P. *Indigenous Agricultural Revolution: Ecology and Food Production in West Africa* (Boulder, CO: Westview Press, 1985).

6. Rostow, W. W. *The Stages of Economic Growth: A Non-Communist Manifesto* (New York: Cambridge University Press, 1990).

OTHER USEFUL READING

• Myers, N., (Ed.) *Gaia: An Atlas of Planet Management* (Garden City, NY: Anchor Press/Doubleday, 1984).

• Scientific American. *Managing Planet Earth: Readings from Scientific American Magazine* (New York: W. H. Freeman, 1989).

• Semple, E. C. *Influences of Geographic Environment* (New York: Holt, Reinhart & Winston, 1911).

CHAPTER 2

The Tropical Climate:
The Great Heat Engine

Climatic types more than anything else set the parameters for land use under different levels of technology. It is important, therefore, to understand in basic terms the defining characteristics of climate and the processes that produce these broad latitudinal zones across the globe. It is necessary to say right away that society's understanding of the working of climate is far from perfect, and this becomes particularly important when we come to look at the phenomenon of climatic change. But we must make the effort to understand the great heat engine that drives all these systems and that allows or constrains different land use types.

The starting point is **solar energy**. In any large geographical area, the input of solar energy determines, or greatly influences, the following variables:

1. The intensity and duration of light available for plant growth
2. The temperature permitting plant growth at different times of the year
3. The wind systems, which move energy from one place to another
4. The ocean currents, which perform much the same function as the wind systems

Incoming solar radiation is moved around by more than just wind and water. The pathways followed by these transporting agents are themselves influenced by the rotation of the globe, deflecting the simple north-south routes that the energy flow would otherwise take. It is not possible in this book to provide a full and definitive study of global climatology, only to outline the main features. If the globe were a perfectly regular smooth and uniform surface, then

17

the climatic zones would be straightforward and would relate closely to latitude north and south of the equator. However, this pattern is greatly disturbed by the irregular distribution of land and water masses in the different hemispheres. In addition, these global climate zones are *locally* modified by the physical characteristics of the land mass over which the air passes. Such characteristics include the following:

1. **The distance from the sea**, so that moisture-bearing winds may be "dried out" by passage over land, leaving less moisture available for precipitation. This is the case, for instance, in West Africa, where the rain-producing winds come in from the sea to the south, and gradually dry out northwards toward the Sahara Desert.

2. **The altitude of the land**, which tends to modify the temperature regime of the air mass. As a broad rule of thumb, it may be said that air cools approximately 1°F for every 300 feet of altitude (1°C per 150 m). Thus the great East African plateau, principally in Kenya, stands around 6000 ft above sea level, so that even on the equator we find a landscape and farming systems that have the appearance of those in the temperate latitudes. In extreme cases, we may even find snow on the equator, as atop Mt. Kilimanjaro in Tanzania. It is often said that as we move upward in the tropics, we replicate the effects of moving toward the poles. It is true that we find rapid changes in vegetation belts as we climb in altitude, say from the semidesert of the Yemen coastal plain to the temperate grassy heights of the escarpment. There is a difference, however, because these higher, cooler areas do not have the longer day-length associated with the more temperate summers.

   Perhaps the best demonstration of altitude in the tropics is to be seen in the Andean ranges of South America. The land there is conventionally divided into the lowland *tierra caliente*, or characteristically hot, moist zone (0 to 3,000 ft); the *tierra templada*, where temperatures are in the range 64 to 76°F (3000 to 6000 ft); the *tierra fria*, which is the grain and potato belt with temperatures between 49 and 64°F (6000 to 9000 ft); and the *paramos*, or alpine meadows stretching to the line of permanent snow with temperatures in the 32 to 49°F range (9000+ ft).

3. **The relief of the land** may well induce heavy rainfall where lower totals would otherwise occur. This is, again, the case on the western slope of the Andes, leaving the eastward-moving air masses much drier when they pass over the mountains, drop their moisture, and reach plateaux or lower country on the lee side. This "rain shadow" effect can produce dramatic changes in climate over relatively short distances. Climate type is therefore the predominant variable in the determination of the naturally occurring vegetation. Climate also influences the processes that establish major soil types, though geology may have a part to play here as well.

Technology plays a role in the adaptation of people to the prevailing climatic type, and this will be examined later in the book. However, at this point, we might mention the existence of irrigation by which water and nutrients are brought into one climatic region from another. Irrigation produces

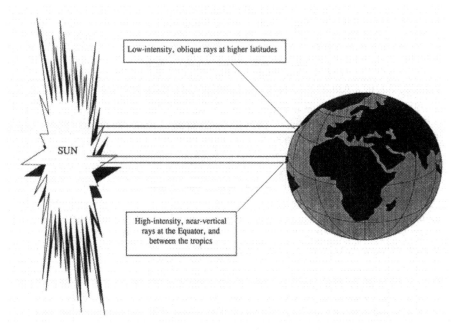

Figure 2.1. The origins of the great heat engine: the differential heating of the earth.

what we might call an **azonal** type of land use allowing, in extreme cases, agricultural cultivation in desert regions such as Egypt. Such activities provide a sort of insulation from the controlling power of climate type over the range of agricultural activities. For most of the people in the tropics, however, the relationship between land use and climate type is a direct and immediate one, and the possibilities for insulation from the climate are minimal. This is why it is so important to understand climate when considering land use.

Pressure Systems and the Tropical Climates

We mentioned in Chapter 1 that energy coming from the sun does not strike a flat surface because the earth is a sphere. Near the equator, and within the tropics generally, the sun's rays strike the earth at a high angle (Figure 2.1), concentrating a greater amount of energy per unit area than is the case to the north and south where the rays strike at a more oblique angle. This means that solar energy is not uniformly distributed over the globe. During the course of a year, the sun moves north and south relative to positions on the earth's surface, being overhead on midsummer's day (June and December) along the tropic lines. Thus, in equatorial regions, the sun passes overhead *twice* during the year. North and south of the tropics the sun is never overhead at any point during the year.

In the intertropical zone is the area of greatest energy receipt. The earth has the capacity to move energy around its surface as well as to dissipate energy back into the upper atmosphere and space. Were this not so, the earth would become unbearably hot and uninhabitable. Despite this movement, there are hotter and cooler regions on the globe resulting directly from differential heating.

The greatest amount of incoming energy is, naturally, in the equatorial belt running through Amazonia, the Congo Basin, Indonesia, and across the intervening oceans. This energy enters as short-wave radiation, but this is not the element that heats up the air. When this incoming energy strikes the earth's surface, some of it is reflected back into the atmosphere and some of it has already been lost because it has been "bounced" off the upper surfaces of clouds. The energy that reaches the ground is absorbed; the degree of absorption depends on (1) the vegetation cover, (2) the reflective index (the albedo) of the terrain (snow reflects a vast amount of energy, which is why it is possible to be sunburned from *below* in such places), and (3) whether the sun's rays strike a terrestrial or aqueous surface. The land surface then heats up, and it, in turn, releases long-wave radiation, which is what contributes most to heating the air near the ground. This is important because it means at least in theory that we could significantly modify climates locally by what we do to change the nature of the surface type and the surface cover.

The tremendous outpouring of energy from the land surface in the equatorial belt (about 5° north and south of the equator) causes the air to heat rapidly and rise. This mass of rising air draws in air from surrounding areas, often ocean masses, which is in turn thrust up into the atmosphere, often as far as 30,000 ft or more. As it rises, this air passes through cooler air and is rapidly chilled. This rapid chilling causes condensation, reducing the amount of moisture that the air can carry and depositing it as heavy rainstorms. Here we have the typical picture of the equatorial belt: hot, steamy, and subject to short, torrential downpours. Rainfall totals in these areas are always high. No month is dry, and conditions for rapid plant growth are always favorable. Because of this, great weights of vegetation may be supported, normally in the form of perennial plants, particularly the trees of the tropical rain forests. These forests, in turn, pump huge quantities of water vapor into the atmosphere, contributing to the conditions favoring heavy rainfall (Figure 2.2). The continuous vast column of rising air in this region results in the creation of characteristic low pressure systems (quite simply, rising air takes the "weight" off the ground below).

But this air cannot go on rising forever, even though it is buoyed up by the arrival of new, heated air from below. As it cools, it becomes denser and has a natural tendency to sink back to earth. On the other hand, it cannot descend in the same zone because it is being uplifted by new air from below. What happens is that the air, once it has risen, tends to "slide over" the upcoming air so that it is pushed north and south of the equatorial belt, falling back to earth around 20 to 30° north and south (Figure 2.3).

It is a characteristic of falling air that, as it falls, it becomes more compressed because it is under increasing pressure from descending air above. As air compresses, it warms. Warm air is able to hold more moisture than cooler

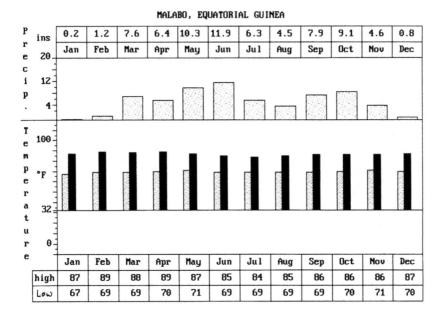

Figure 2.2. A characteristically equatorial climate. (Map courtesy of PC Globe, Inc., Tempe, Arizona, U.S.A. Copyright 1991.)

air, so it is unlikely that these descending air masses will shed any of their moisture. Thus, in these latitudes we find descending, stable, dry air masses associated with high pressure systems (due to the "weight" of falling air). It is this phenomenon that gives rise to most of the great hot deserts of the world, best exemplified by the Sahara. These hot deserts are better developed in the northern hemisphere because of the great breadth of the land masses there. In the southern hemisphere, the narrowing of the continents brings the influence of the oceans into play in these latitudes, and deserts do not develop to the same extent (except in Australia). Mostly we find narrow fringe deserts such as the Atacama of Chile and the Namib of southern Africa. These very dry conditions do not favor human settlement or agriculture to any significant degree, unless there is an **exogenous** or **azonal** factor such as a mountain range or the possibility for irrigation.

Now we come to the third stage of the intertropical climatic cycle. Back at the equator there persists the insatiable demand for air caused by the need to fuel the continuous rising-air column. It will be recalled that this is associated with a low pressure system. The descending air to the north and south creates a high pressure system, and air naturally moves from high pressure to low. Thus the air is drawn in from the desert regions or oceans and recycled through the equator. This continuous process is described as a "cell", and we find these cells developed north and south of the equator, as shown in Figure 2.3. They give us our two extreme types: the equatorial rain forest and the subtropical

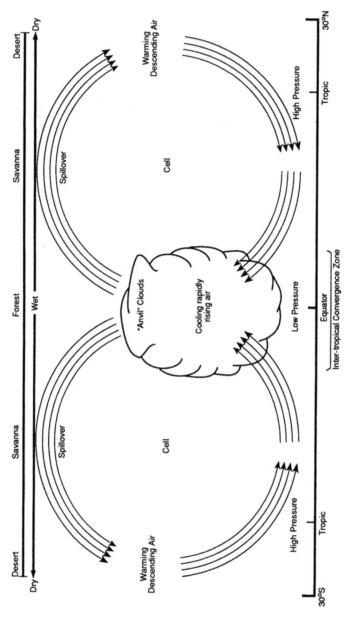

Figure 2.3. Pressure system and cells.

deserts. Both of these areas are characterized by relatively low densities of population. From the perspective of land use, our attention focuses much more on the area **between** the two, which is often associated with a **monsoon**, or tropical continental, type of climate. From the agricultural point of view, there is not much difference between these types, though they are often differentiated by geographers because of the different factors that bring about the juxtaposition of high and low pressure systems.

The Tropical Continental and Monsoon Types

Simply stated these two systems are those that have a strongly seasonal rainfall pattern bringing moisture sufficient for some form of unirrigated agriculture. To understand their formation, we must return to the fact that the sun annually moves north and south relative to fixed points on the earth's surface. It will be recalled that the equatorial low pressure system draws in air from the north and south. Where this air comes together and is uplifted, it produces a phenomenon known as the **Intertropical convergence zone** (ITCZ). This narrow zone, extending about 5° north and south of the equator, has its natural midpoint a little north of the natural equator.

During the year, the ITCZ itself moves north and south following the highest angle of the sun's rays. This movement is shown, for the continent of Africa, in Figure 2.4. During the northern summer, the ITCZ moves northward and so the winds fueling it from the south are also dragged northward. When this happens, air masses coming from the mid- and south Atlantic are pulled in further and further over the West African coast toward the belt known as the **Sahel**, which lies between the wet coastal hinterland and the arid Sahara Desert. As this moist air is pulled inland, it is subject to considerable heating near the ground, and the lower part of the air mass becomes unstable and has a tendency to rise. Once more, a rising air mass is cooled and produces the precipitation that fuels agricultural growth. So we need two things for precipitation to happen: (1) a moist air mass and (2) a "triggering" mechanism to cause instability and produce the rain. This seasonal reversal of winds produces the characteristic climate type over vast and densely populated regions of the tropics such as the North Indian plains and Ethiopia.

As the sun moves south, the winds are less and less able to penetrate the northern continental masses and in their place the dry air masses from the high pressure systems take over. So in West Africa, for instance, there is a *reversal* of wind systems during the year, from the southwesterly monsoon to the dry northeasterly *Harmattan*. In the southern hemisphere, again because of the narrowing of the continents, there is a much less developed form of this climate.

The typical Asian monsoon climate produces much the same pattern of rainfall, but is caused by the differential heating of land and ocean. During the summer, the land is hotter and draws in moist air from the sea. In winter, the reverse occurs and dry conditions result. Perhaps the best-known monsoon area

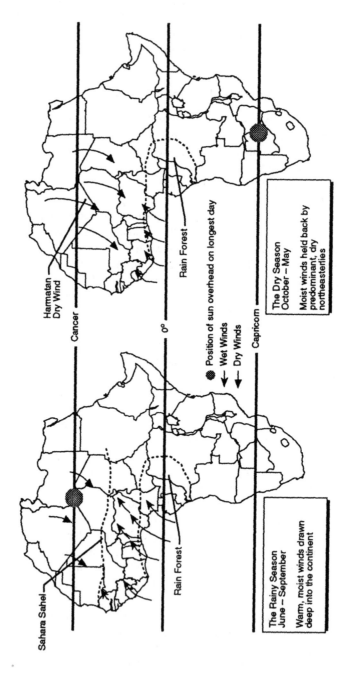

Figure 2.4. The West-African monsoon.

is in India, where only 5% of rainfall around the Bombay area falls outside the four summer months. It should be pointed out that both the tropical continental and monsoon types of climate are often referred to as monsoon types because of the similarity of effect. It is easier if we refer to them both as monsoons.

These two climates, then, are typified by a markedly *seasonal* rainfall pattern (see Figure 2.5), and the land-use types have to adjust to this constraint. In addition, the further one moves north or south of the equator, the *shorter* becomes the rainfall season, the *lower* the totals of precipitation, and the *narrower* the range of crop types that may be grown in this belt.

Climate Constraints

In the temperate areas, plant growth is seasonal. However, the primary constraint is temperature. Plants generally become inactive at temperatures below 42°F. In the intertropical zone, temperature is rarely if ever a constraint since there is sufficient warmth throughout the year for plant growth. Indeed the world's highest temperatures are to be found in this zone, usually in the desert regions where solar energy is unobstructed by cloud cover. Where moisture is available, it is possible to grow plants throughout the year and often to produce more than one crop from the same piece of land. In the humid tropics — the equatorial and subequatorial belts — the yields are high indeed. The maximum rate of dry matter production in the world comes from sugar cultivated in Hawaii (albeit with the benefits of modern agricultural methods), yielding more calories per acre than any other crop. This seems odd when we consider the concentration of poverty and malnutrition in the tropics, but that is mostly associated with high densities of population in areas of unimproved seasonal agriculture. Perennial grasses yield about 24 tons per acre in the tropics, compared with about 8 tons per acre in the temperate zones.

Related to temperature is energy. In fact the total number of daylight hours varies remarkably little across the surface of the globe. The greater intensity of daylight in the equatorial areas is offset by cloud cover and the shortness of the day (never much more than 12 hr). The real advantage of the tropics is in the humid areas where perennial crops may be grown because of the year-round growing season. Ironically, the area of the globe with the greatest amount of available solar energy to fuel plant growth is in the desert regions, where almost nothing grows because of the near-total lack of precipitation.

Precipitation, or more strictly moisture availability, is the principal constraining factor on land use in the tropics. This is, of course, not true of the humid tropics. There the *intensity* of rainfall can be exceedingly damaging if the form of land use is not adapted to that characteristic, since it will tend to leach the nutrients from the soil and cause massive erosion if not checked. Rainfall exceeding 0.4 in./hr is likely to be damaging unless controlled by vegetation cover, and totals of this sort are common in the humid tropics.

In the areas of seasonal rainfall, moisture availability becomes the constraining factor in several ways. First of all the range of crop types will depend on the amount of precipitation, its intensity, and the length of the rainy season.

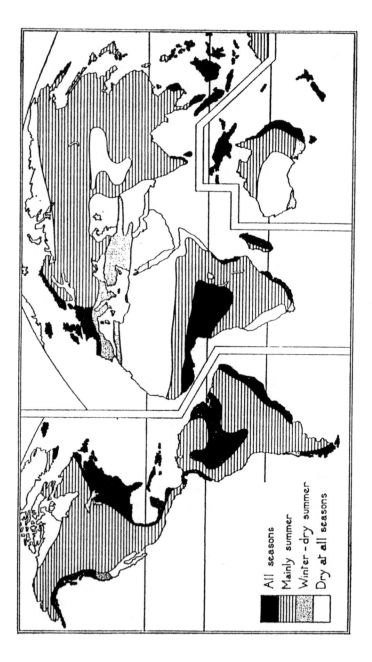

Figure 2.5. The seasonality of precipitation.

We have noted that as we move away from the equator, the rainfall totals diminish and the length of the rainy season shortens. However, there is an additional factor. As the rainfall totals diminish, the coefficient of variability rises (Figure 2.6). This means that not only is there less moisture available, but the chance of *not receiving* that total increases. To put this another way, the range of crops that may be grown diminishes, reducing the farmers' options to hedge their bets and increasing the risk of their failing to get a good harvest from these crops. So agriculture and pastoralism (subsistence livestock raising) become increasingly risky activities. These are, indeed, the parts of the world familiarly associated with drought: the Sahel and, in the past at least, monsoon India. The land-use system has to find ways of coping with this risk if people are to survive over time in these regions. We shall see later in this book that societies did just that, though often under less severe population pressure than currently exists.

Though it is not usually considered in a section on something as broad as climate, some mention is needed of wind in a book dealing with land use. The seasonal areas often experience a reversal of winds, and during the dry season these can bring a powerful desiccating influence across the land. If soil moisture is not effectively conserved at such times, the enormous excess of evaporation and transpiration (the passage of moisture by plants), relative to moisture availability, will parch the land. Furthermore, winds in the drier areas have a strong erosive power and are known to have stripped away dried-out topsoil in many parts of the tropics, such as Mali.

Climate and Climate-related Zones

Reference has already been made to the humid and dry tropics. These are the two extremes. Between these two is a gradation of types usually named after their characteristic vegetation types. Most commonly, these are types of wood and savanna (called variously *llanos* or *campos* in Latin America or *veldt* in southern Africa). It has been customary to delineate tropical areas according to their climax vegetation forms. These are the main assemblages of vegetation that would occur if people were not interfering in the system by cultivating, clearing, burning, grazing, and such other activities that modify the natural situation. Thus burning is thought to have created the perennial grasslands of Africa, which, all things being equal, should probably have been wooded. This type of vegetation maintained by human practice may be termed a **subclimax** form and is dependent for its stability on the maintenance of the practice. Sometimes this climax — or "end form" — vegetation is also modified by underlying geology and other factors. The problem in many places is to determine what the climax form would be because centuries of environmental modification have taken place as a result of human occupation. A broad classification of natural vegetation types is, however, presented in Figure 2.7.

In association with the equatorial type of climate, the tropical rain forest is dominated by perennial species and bears a tremendous weight of vegetation,

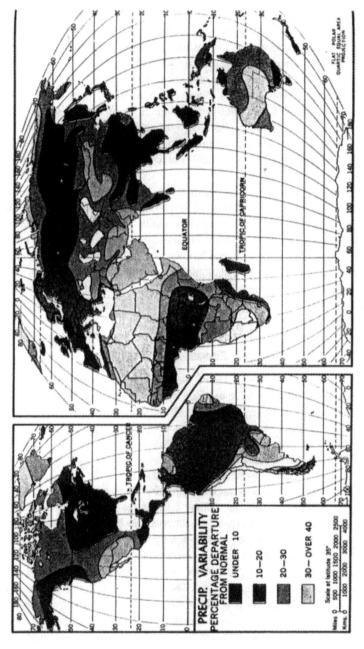

Figure 2.6. Precipitation totals and variability. (From Trewartha, G. *Elements of Geography*, McGraw-Hill, 1957. With permission.)

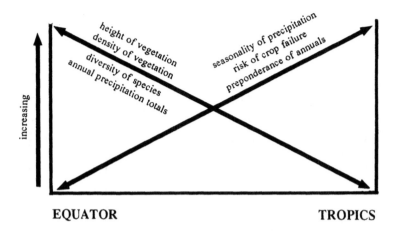

height of vegetation
density of vegetation
diversity of species
annual precipitation totals

seasonality of precipitation
risk of crop failure
preponderance of annuals

increasing

EQUATOR   TROPICS

Figure 2.7. Environmental changes with latitude.

often forming a closed canopy, or several canopies, high above the ground (100 to 200 ft and more in places). This vegetation type is associated with an enormous range of species randomly scattered through the forest. Growth is continuous and rapid, and plants may be found at every stage of their reproductive processes at any time. The closed canopy protects the land surface from the ravages of intense downpours, though the small amount of light penetrating the canopy does not encourage a dense ground cover, despite the images produced in Tarzan movies. The tropical rain forest is, in ecological terms, the most *productive* and *diverse* ecoregion (or **biome**) on earth.

As we move away from the equatorial belt, variations in the climate produce several characteristic changes in the vegetation type (Figure 2.8). First, the height of vegetation diminishes along with the competition of plants for light. Forests give way to woods, to parks, to wooded savannas, to bush, and eventually, to semidesert. Second, the weight of vegetation per unit-area diminishes, and year-round species survive only through special adaptations of site, or by having deep roots, for instance. Third, the number of species diminishes as the climatic constraint increases. Fourth, the perennials yield to annuals, reflecting the seasonality of rainfall, and eventually the annuals give way to **xerophytic** (dry-adapted), and **adventitious** (able to respond to occasional, rare rainstorms) forms of vegetation.

And so the scene for a study of tropical environment and land use is set by these broad, defining characteristics of climate. Of course, other factors will come into play locally, but in general, when we come to look at the land-use systems that evolved in the tropics, we shall find compelling adjustments to the realities of climate.

The Question of Climate Change

Everything we have said so far in this chapter assumes that the climate is a stable entity, and we have considered only the spatial changes: how it varies

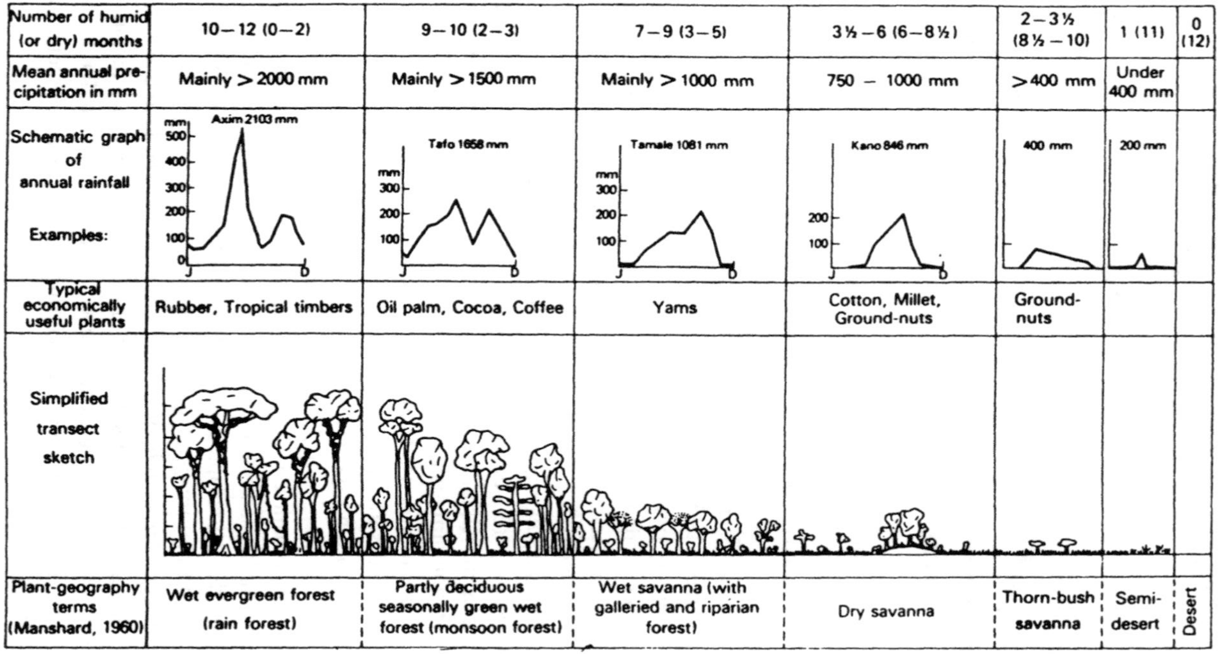

Number of humid (or dry) months	10–12 (0–2)	9–10 (2–3)	7–9 (3–5)	3½–6 (6–8½)	2–3½ (8½–10)	1 (11)	0 (12)
Mean annual precipitation in mm	Mainly > 2000 mm	Mainly > 1500 mm	Mainly > 1000 mm	750 – 1000 mm	> 400 mm	Under 400 mm	
Schematic graph of annual rainfall	Axim 2103 mm	Tafo 1658 mm	Tamale 1081 mm	Kano 846 mm	400 mm	200 mm	
Examples:							
Typical economically useful plants	Rubber, Tropical timbers	Oil palm, Cocoa, Coffee	Yams	Cotton, Millet, Ground-nuts	Ground-nuts		
Simplified transect sketch							
Plant-geography terms (Manshard, 1960)	Wet evergreen forest (rain forest)	Partly deciduous seasonally green wet forest (monsoon forest)	Wet savanna (with galleried and riparian forest)	Dry savanna	Thorn-bush savanna	Semi-desert	Desert

Figure 2.8. An ecological transect from equator to tropic. (From Manshard, W. *Tropical Agriculture: A Geographical Introduction*, Longman Group U.K., 1974. With permission.)

from one place to another on the globe, and how this lays the foundations for land-use systems. There is, however, a complication, namely: "What if the climate in general is subject to change?" How would that change affect the land-use possibilities in the tropical area?

We know for a fact that climate has changed in the past. There have been at least four great ice ages during which massive changes occurred across the globe. We also know that in Europe it was common, in Elizabethan times, to schedule fairs on London's frozen River Thames; something that would be impossible now. In the tropics, we have fossil dunes in moister areas (Nigeria), evidence of changing lake levels (Chad), and other phenomena that *might* indicate climatic change. However, we need to be clear right away that we can *prove* almost nothing with regard to *contemporary* climatic change. Our understanding of the forces that shape our climate is in its infancy, and our ability to predict is uncertain. Furthermore, much of the work is conducted in the temperate areas and we have to extrapolate from these to the tropics, adding more uncertainty to the equation. Still, this is something we need to follow closely.

Why would climatic change be particularly critical in the tropics? The first point of importance is that in the tropical regions, the majority of people derive their living *directly* from the land. Since their systems of cultivation are directly linked to the weather, they have little ability to insulate themselves from, or ensure themselves against, fluctuations or more systemic change. Thus, if the climate changes, it has the potential to directly affect a great many people in the area of subsistence. In addition, a great many people live in areas — such as the monsoonal zones or areas of highly seasonal rainfall — where relatively small fluctuations in the total rainfall, its onset or its duration, could be critical. In marginal areas, a downward trend could render these places uncultivable, as along the southern margin of the Sahara. For instance, if anything influenced the amount of solar energy reaching the equatorial area, it would reduce the northern and southern movement of the ITCZ and its associated monsoonal winds, encouraging the desert margins to expand into the area of cultivation. This, some argue, is what is already happening, as we shall see when we come to look at desertification (Chapter 15).

There are two types of climatic change that concern us, though we shall have to treat the subject quite briefly. The first type of change is a global atmospheric change, meaning that some modification takes place in the entire heat engine of the gaseous envelope around our globe. Many reasons are hypothesized for such a change, including "greenhouse" gases, which would trap additional amounts of the outgoing long-wave radiation and lead to a gradual warming of our atmosphere. The scientific community, while it cannot in a strict sense *prove* any such thing, shows an increasing acceptance of the probability that this may be happening. As far as the tropics is concerned, the destruction of the tropical rain forest is the only area where there is a speculation of causation between human activity and global climatic change. Proponents of this argument state that eliminating these vast, forested areas reduces the ability of a part of the world to convert carbon dioxide (CO_2) to oxygen (which trees do as part of their normal life process), thus contributing to the buildup of CO_2, thought to be one of the "greenhouse" gases. As we have mentioned, this is not a quantitatively and inevitable *proven* consequence.

The United Nations Intergovernmental Panel on Climatic Change, reporting in June 1990, gave as their best estimate that, by the year 2030, the tropical rain forest areas would show no significant change in climate, while in West Africa and southern Asia, there would actually be an increase in precipitation — though their model does not reveal variations within these vast regions.[1] So, at present, there is little in the global atmospheric change modeling to suggest alarm for the areas that we are considering.

The second type of climatic change, equally hypothetical, may be referred to as "local" modification of climate, though it can extend over a considerable area. This involves not changes in the composition of the global atmosphere, but changes in the conditions affecting air masses over specific regions, nearly always induced by land-use activities. Two of these may be mentioned briefly.

Text Box 2.1
Estimates for Climatic Change by the Year 2030

Southern Asia (5 to 30°N, 70 to 105°E)
The warming varies from 1 to 2°C throughout the year. Precipitation changes little in winter, and generally increases throughout the region by 5 to 15% in summer. Summer soil moisture increases 5 to 10%.

The Sahel (West Africa) (10 to 20°N, 20°W to 40°E)
The warming ranges from 1 to 3°C. Area mean precipitation increases and the area mean soil moisture decreases marginally in the summer. However, throughout the region, there are areas showing an increase and decrease in both parameters.

In the first case, clearing of the land for the cultivation of, for example, cotton in a monsoonal area would tend to change the reflective index of the land (the albedo). Anyone familiar with black-and-white aerial images of land use will know that these cleared areas of land appear bright on the photographs. At the beginning of the rainy season, when the moisture-laden air is passing over the land that has been cleared, that land is not absorbing as much sunshine as it did formerly. Consequently it is less heated and is less able to heat the air passing over it or to induce or "trigger" the air to rise and produce the rain. So the monsoon may well fail.

The second hypothesized case is that of dust. Again, through the clearing of the land, soil surfaces are left bare and desiccated. Wind is then able to pick up the finer particles and carry them into the upper atmosphere. Indeed, particles of this sort have been identified from West Africa as far away as the Caribbean following the drought of 1968.[2] This dust layer, being relatively dark and dense, absorbs the energy of the sun passing through it and becomes a band of higher temperatures in the atmosphere. So, any air mass that has been heated near the ground, and that is rising to cool and produce rain, is suddenly going to find itself passing through a band of much higher temperatures that could prevent the air mass from rising any further, which will again prevent it from producing any rain.

Once more, it is necessary to stress that although we can observe dust in the atmosphere resulting from changed land use, it is still *speculative* to connect this in any causative way with long periods of drought occurring in West Africa since 1968. On the other hand, this is an area demanding continuing research.

Broad Divisions of the Tropical Realm

Using the variations in climate, it is now possible to divide the tropical and subtropical areas into **ecoregions** because "at the broad level [climate has an] overriding effect on the composition and productivity of ecosystems."[3] These collections of ecosystems, derived from the work of Bailey, represent characteristic zonal assemblages of plants based on climate and altitude. The shading of each zone shows the dominant climate type, while the name signifies the principal vegetation type. It should be remembered that each zone, unless a sharp break of altitude occurs, merges into the next along a line of transition, rather than following a clear line of demarcation. The broad zones represent the natural adjustment to the parameters of moisture availability, relief, and altitude and present us with those forms of life that are best adjusted to the opportunities and constraints for life activities in different places. A simplified form of Bailey's map is illustrated as Figures 2.9, 2.10, and 2.11 and will be used as the reference point for the discussions in the rest of the book.

Bailey subdivides the humid domain into "(a) the savannas or alternating wet and dry tropics, and (b) the rain forest or wet tropics, on the basis of seasonality of rainfall, total annual rainfall, and the density of plant cover."[3] Meanwhile, the dry tropics "comprises the arid and semi-arid regions . . . and has discontinuous vegetation of steppe, xerophytic bush, and desert, types with only intermittent runoff."

Summary of Climatic Influences

Since we are interested primarily in the way climate influences human landusing activities at different levels of technological advancement, it is useful to summarize the main constraints imposed by what we have observed here.

1. **Seasonality** imposes a limit on societies in areas of seasonal moisture availability, since it shortens the growing season, and introduces the risk of crop failure with lower and more variable rainfall totals and the need to compress labor activities into bursts of frantic energy. It also means that the agricultural system has to carry the population through a "hungry season" when plant growth is dormant.

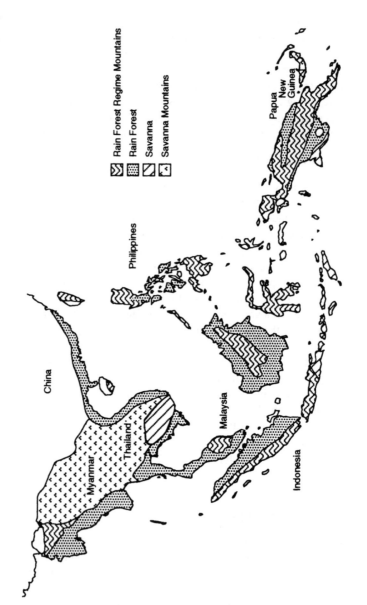

Figure 2.9. Ecoregions of the world. (Modified from Bailey, R. G. *Environ. Conserv.*, 16(4): suppl. [1989].)

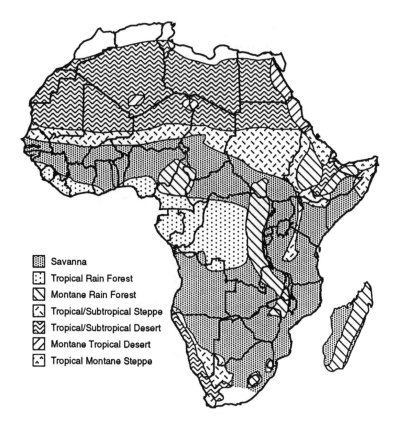

Savanna
Tropical Rain Forest
Montane Rain Forest
Tropical/Subtropical Steppe
Tropical/Subtropical Desert
Montane Tropical Desert
Tropical Montane Steppe

Figure 2.10. Ecoregions of the world. (Modified from Bailey, R. G. *Environ. Conserv.*, 16(4): suppl. [1989].)

2. **Temperature**, although not the principal constraint in the tropical areas, when high generally accelerates the biochemical processes and results in a great deal of moisture being evaporated or transpired back into the atmosphere. Thus the breakdown of organic matter is rapid, and nutrients may be lost to the system if plants are not adapted to rapid recycling. Plants in these areas may have to be more moisture-demanding than their counterparts in cooler climes.

3. **Humidity** may make the storage of crops during the hungry season (the long, dry period) something of a problem because it encourages the growth of molds and mildews, leading to losses in storage that have been estimated as high as 30%.

4. **Intensity** of rainfall means that a vegetative cover is an important instrument in protecting the ground against the ravages of erosion.

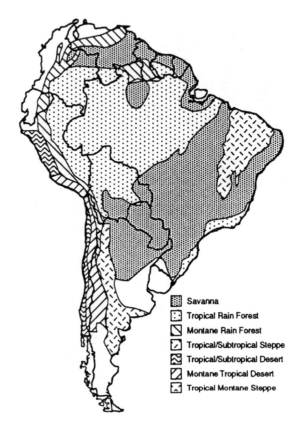

Savanna
Tropical Rain Forest
Montane Rain Forest
Tropical/Subtropical Steppe
Tropical/Subtropical Desert
Montane Tropical Desert
Tropical Montane Steppe

Figure 2.11.   Ecoregions of the world. (Modified from Bailey, R. G. *Environ. Conserv.,* 16(4): suppl. [1989].)

REFERENCES

1. United Nations. *Intergovernmental Panel on Climatic Change: Summary Report,* New York: WMO/UNEP, (1990).
2. Rapp, A. "A Review of Desertization in Africa," SIES Report No.1, Swedish Natural Science Research Council, Stockholm (1974).
3. Bailey, R. G. "Explanatory Supplement to Ecoregions Map of the Continents," *Environ. Conserv.* 16(4):307-9 (1989).

OTHER USEFUL READING

• Manshard, W. *Tropical Agriculture: A Geographical Introduction* (London: Longmans, 1974).
• Navarra, J. *Atmosphere, Weather and Climate: An Introduction to Meteorology* (Philadelphia: W. B. Saunders, 1979).
• Ramage, C. S. "El Niño." *Sci. Am.* 254(6):76 (1986).

2. The high temperatures and humidity accelerate the processes of decomposition, leaving the soils poor in nitrogen and phosphorous.
3. Leaching (the washing of minerals and organic matter through the soil) and erosion may result from heavy downpours. In general the nitrogen and organic matter content of the soil varies inversely with the temperature. Above 70°F (21°C) the decomposition of the humus layer proceeds at a higher rate than its formation.
4. There is an inability of all but the very exceptional soils to build up any store of organic matter to act as a reserve of fertility.

Throughout many parts of the tropics, **lateritic** soils have formed creating a hard, impermeable layer or "pan" that rings like steel if struck with a metal instrument. These laterites result from the steady leaching of plant foods and bases, leaving concentrations of sesquioxides of iron and hydroxides of aluminum that form concretions. This process produces the typical red color of many tropical soils. In general tropical soils have a shallow organic layer of just an inch or so. Much of the organic material is tied up in the vegetation cover, the removal of which leads to rapid sterilization of the soils. These soils are deep and largely unstructured. Most of what we may observe by looking at a vertical section of the "soil" is just weathered inorganic material. Over considerable extents of the humid tropics, we find soils derived from the basement rock, which is often granitic, resulting in sandy soils through which nutrients may be leached with ease. The lesson is that it is important to recognize the repository of energy represented by the vegetation cover and the critical role of the vegetation in maintaining soil fertility and stability. Both soil and vegetation are intimately linked in a cyclical process. Removal of the vegetation cover requires that some adjustment be made for the loss of this cycling mechanism and for the protection the cover had traditionally provided against erosion and leaching. This has been a hard lesson for many agricultural "modernizers" to learn. Traditional systems, on the other hand, have incorporated many methods for copying the natural processes and maintaining stability under different levels of population pressure.

The basis of many traditional systems has been the successful local recycling of energy by harvesting the natural vegetation and allowing for a **fully regenerative fallow**. Though it is true that the early systems did not import large amounts of energy across farm boundaries, as is characteristic of "modern" systems, nor did they *export* large amounts as commodities either — which is why we refer to them as **locally recycling** their energy. Occasionally, energy from a wider area was brought into the farm and concentrated on the area of production. For instance, where population has grown, societies have shown a capacity for strengthening the soil by the application of night soil and domestic waste (as seen in the concentric zones of decreasing intensity of cultivation around the towns of Northern Nigeria) or by techniques such as "green manuring", whereby crop residues are dug back into the soil to maintain fertility. Generally, these intensification techniques are more *labor demanding* than those that simply recycle the energy in the immediate system, and societies show a tendency to opt for the most stable, labor-efficient system unless population pressure demands otherwise. Let nature do the work.

There are in the tropics, however, local exceptions to the rule of generalized low soil fertility, and these are often associated with high densities of population and the development of civilizations of a more settled, differentiated, and monumental type. The first of these exceptions is produced by *irrigation*. Irrigation is a method of capital formation by which societies, where physical circumstances permit, channel water from a broad catchment into a concentrated area of cultivation. This can result in the movement of water across high mountain ranges (as the *Inca* were able to do in the Andes) or for tens of miles under the desert (as was achieved in the *qanat* system of desert Iran).

It is important to remember that the water is also an agent of transportation for nutrients from a broad area, and it is this that provides for the annual renewal of growth media. In fact this is almost a form of *hydroponics* (cultivation in water), because the growth owes little to the soil *in situ,* which is merely an anchoring mechanism. Clearly, the opportunities for irrigation are severely constrained by topography and the availability of a water source. However, these fortunate areas have provided the opportunity for continuous cultivation over almost unimaginable periods of time. Perhaps the best known among such systems is the Nile in Egypt, which has supported civilization there for almost as long as agriculture has been practiced. Each year the rainfall in Uganda and Ethiopia charges the headwaters of the Nile with an increased flow and renews the nutrients in suspension and solution. The river in flood brings a ribbon of opportunity stretching through the desert to the otherwise arid land of Egypt. In the past the river would flood regularly into prepared basins, and the cultivators would secure one or more crops from the rich alluvial mud. This allowed for some regularity in the surplus of food, and the possibility to support nonagricultural people, without which there would have been no temples at Karnak, no pyramids at Gizeh.

Related to irrigation is the occurrence of alluvial areas such as great river deltas, which are the repository for vast amounts of mud and nutrients brought down by the rivers (a form of natural irrigation). Such deltas in Asia (the Mekong for example) are able to support huge concentrations of population on unimproved agriculture. The principles at work are the same as those in Egypt, though there may not be the same degree of hydraulic engineering required to capture the benefits. Associated with irrigation in Asia is rice, which is the most productive of grain crops, and this crop in turn allows great population densities to be supported. Generally it is impossible to have irrigation systems without sophisticated systems of social control and regulation, which is why these areas were so often, in ancient times, associated with the development of civilization. Indeed, it is true to say that much of mathematics and geometry found its beginnings in the need to divide up land and apportion water in Mesopotamia. (The Sumerian base-60 system of mathematics gave us our crazy division of the clock.)

There existed another interesting system known as **water harvesting**. In this case societies clear vegetation from the surrounding hillsides to encourage runoff, and this was deflected by barriers into defined channels and concentrated in lower-lying areas of cultivation. Water harvesting was practiced by the Nabateans in what is now Israel and the northern Hejaz area of what is now Saudi Arabia and allowed for the development of extraordinary concentrations of people into towns. Briefly, irrigation spreads the growing season, allows cultivation in otherwise uncultivable areas, broadens the range of crops, and reduces the risk of crop failure. The high densities of population it can support also facilitate the transportation and marketing of surpluses, in turn favoring the division of labor. So, while it may not be strictly correct to talk about irrigation under the heading of "soils", irrigation does provide an important azonal exception to the broad patterns and qualities of tropical soils.

The second major azonal exception concerns the occurrence of local geological factors that modify the general circumstances. The principal example is the distribution of volcanic soils. In these instances we *do* find depths of mineralized material that form a store of nutrients for growth provided by the decomposition of the weathered rock. Most often this soil type is associated with mountainous terrain because of its old vulcanicity. Across the tropics we see numerous examples of these high-density montane high-altitude systems as in Central America, Cameroon, Rwanda, Java, and the Yemen. Sometimes these systems are also associated with irrigated terracing (Philippines) or with unirrigated terraces (Rwanda). Often great densities of population developed in these fertile regions, too, and expanded under pressure into the surrounding areas, though the people who left may not have been able to practice the same systems of agriculture. The volcanic highlands became the "hearths" of entire peoples and races, from which they spread over much of the earth. In this way, for instance, the Arabs spread from the Yemen into Arabia, and the Bantu from upland West Africa over the center, east, and south of that continent.

In general, traditional societies depended on the locally occurring sources of fertility and were unable to do much to modify that. The development of "modern" farming has been the story of bringing concentrations of fossil and "processed" energy to bear as a supplement to the availability of local naturally occurring nutrients. Mention should also be made of the fact that in the drier tropics particularly, soil formation is a very slow process (despite the rapidity of biochemical processes). Some estimates for the semiarid place soil formation at a rate of 1 cm per 100 years.

The Biological Environment

It is not only plants in the tropics that show a high degree of species diversity; the same is true for fauna as well. This is of significance to our study in several ways. First of all, both stored crops and crops in the fields are constantly at risk from a broad spectrum of birds, reptiles, and insects. A typical farm in the tropics features a child somewhere whose job it is to frighten away predators from cattle herds or scare off the birds that constantly prey on the crops. Traditionally farmers have tried to combat this broad spectrum of pests and diseases by opting for diversity wherever possible, rather than concentrating risk into a monoculture (though rice is an exception to this because of its unique suitability for irrigation).

Some mention should also be made about the perceived "unhealthy" nature of the tropics, so heavily emphasized by Gourou, for instance. Relatively recent history is filled with depressing tales of epidemic and endemic diseases, as well as enzootic and epizootic [animal] ones also. We have to get a perspective on this. It is true that the population of the tropics, and to some extent their crops and animals, is distinctly unhealthy compared to the population of temperate lands. There is a general belief that somehow this is the "natural order of things." People are weakened by malaria, parasites spread by communal living and shared water supplies (bilharzia), and exposure to debilitating waves of sickness. This reduces the ability of the people to work, reduces their productivity, and reduces their ability to fight off further attacks of disease. It is then nearly impossible for them to work any harder to improve their lot (the vicious cycle of poverty). There is, no doubt, great truth in this assertion, particularly in the humid tropics. On some occasions whole populations of animals have been decimated (the rinderpest epizootic among the Maasai cattle herds in Kenya at the turn of the century), and their human counterparts have suffered a similar fate due to such illnesses as smallpox and influenza (Kenya again). However, these were sometimes *introduced* diseases in areas where people had no locally developed resistance. For the same reason, in reverse, West Africa was known as the "white man's grave" because Europeans had little immunity to the diseases endemic there, and had a habit of dying.

But there is another dimension to this story. It is true that suffering, morbidity, infant mortality, etc. are high in the tropics. But is this a regional characteristic of "being tropical"? We should remember that a few hundred years ago the temperate lands showed the same characteristics. It is only relatively recently that smallpox, typhoid, cholera, malaria, and diphtheria have ceased to exact their toll in the West. The improvement has been a result of better public health and improved access to preventive and curative medicine. Most of all it is due to a higher standard of living for the mass of people and better general education levels. Such social characteristics and facilities are rudimentary in the tropics, and this, rather than an endemic, ordained physical condition, explains the generally "unhealthy" nature of the tropics. Indeed, and perversely, there is now a worry about the dramatic rise in population growth in the tropics resulting from better health care and nutrition and the fact that more people stay alive and live longer. The biotic environment remains "unimproved"and this goes along with a heavy toll on the human and animal population. Across large parts of Africa, it is all but impossible to keep cattle because of their susceptibility to trypanosomiasis (*nagana*, or sleeping sickness), which is spread by the tsetse fly (*glossina*). It is, therefore, not possible to achieve the benefits of mixed cultivation or to use ox-drawn implements for farming. As a result, people often lack the resources to fight the attacks of diseases and pests, except through the store of ethnoscience, which is now rapidly being lost and which for so long was disdained as "folk medicine". For instance, the burning of grasslands, an action that so troubled the colonial powers, was often conducted at the onset of the rains to eradicate the tick population that spread bovine East Coast Fever in Africa. The burning also brought on a flush of nitrogen-rich grasses.

It seems, then, that just as we have come to accept poverty as a "natural" condition in these areas, so we accept that they are inherently unhealthy. In fact neither condition may be natural at all, and the former may well be the best explanation of the latter. Rather than saying that "people are poor because they are sick," it might be more correct to say that "they are sick because they are poor." Consider, for instance, the enormous difference between the white and black populations in the American South. The black population was typified by a host of debilitating diseases such as tuberculosis, pellagra, and malaria, in sharp contrast to their white neighbors. This could not have been because of the natural environment per se, but because of their greater susceptibility to disease through their reduced economic condition.

Here we come across a phenomenon that bedevils our understanding of the "tropical condition". Quite often popular mythology results in what is a reversal of cause and effect. In much of the colonial literature, the African farmer is depicted as "lazy". This universal social attribute, according to the Western observer, accounted for their low levels of productivity and their general failure to achieve the modernization of the West. As we saw earlier, the determinists blamed this racial fatigue on the climate. Laziness may well have resulted from the fact that, for part of the year, it is impossible to farm because of the absence of rain. Also, the farmer may well be carrying malaria, a condition that is exacerbated by the onset of rainy conditions, at the very time when the most work is needed to prepare the fields for cultivation. Once more, however, we come back to the questions: "Why is this part of the world so poor? And, why is it that the temperate countries achieved the development that allowed them to become healthier people?" It may well be that the progress of the temperate areas was achieved at the expense of the tropics, and this, in turn, prevented the tropical areas from moving along the curve of social and economic advancement. This is one of the great imponderables in any study such as this, and it is always much more comfortable to avoid the question totally and concentrate instead on the problem in technical terms, thereby avoiding one of the great truths. At the same time, as mentioned earlier, this book is not about apportioning blame; it is about understanding. Until we understand causation, we are unlikely to be able to do much about devising remedies. The real issue is to devise sustainable solutions and to consider the means of effecting them.

Economic Constraints

When we look at "modern" farming in the West, the main characteristic that strikes us at once is the enormous amount of capital involved. We shall look at this in more detail when we consider the question of energy in the next chapter. Most of the traditional farmers bring to bear only two factors of production on their agricultural systems: land and labor. In the West we have to add the dimension of capital manifested by machinery, infrastructure, the development of science and technology, fossil energy, etc. In the tropics, the use of capital is minimal, and production is constrained principally by the limits of human and animal labor. The average farm family acting alone can manage about 2.5 acres, and this allows for relatively small and often irregular surpluses, affording little opportunity for capital accumulation. In addition, inaccessibility makes the marketing of this surplus difficult, and storage problems prevent the bulking and saving of supplies. It is salutary to note that ancient Egypt was rarely able to produce much more than a 5 to 10% surplus to support its infrastructure. The traditional farmer may be said to be operating in a "precapital" system of production, and many "developers" target the capital gap as the principal constraint on moving these systems to higher levels of productivity.

Some capital formation did exist in the precapital systems. Where societies chose to terrace hillsides to extend their cultivation, or where they invested labor in irrigation networks, they were able to secure a flow of benefits beyond the present harvest. However, these were the exception rather than the rule, and they required further great outpourings of *human labor,* which is why they were sometimes associated with slavery. It is fair to say that the economic constraint is the principal one at present, since it prevents the capitalization of farming that is necessary for increased productivity. Once more this brings us back to the poverty of the tropics, rather than to any inherent natural constraint on advancement. The economic dimension will be considered in the context of the historical process and of the case studies later in this book.

Social and Cultural Factors

Here we enter another great area of mythology in much of the evolution of Western historical orthodoxy about the tropics. Report after report during the 1960s and 1970s used to be filled with a disclaimer about "social factors" that caused projects to fail. Generally, traditional societies were seen as a barrier to their own advancement by outsiders. This was well described by Caroline Hutton and Robin Cohen in their classic article on "Obstacle Man".[1] Here we have to carefully examine our use of the word "traditional", which has been somewhat loosely used in this text so far. Tradition is that body of accepted lore within a group that guides its behavior. Writers in colonial times, and indeed some today, speak of tradition as though it is the mindless, ritualistic determinant of behavior, making tradition an *independent* variable. The people, according to this view of tradition, do what they do because they know nothing else, and they are controlled by the "dead hand of tradition". Indeed these societies were characterized by great continuity in their behavioral practices and by the encoding of behavior into systems of social control of long standing. To some extent these practices may have been ritualized, but doing so helped people learn them, recognize them, identify with them, and accept them as belonging to the group. This was, after all, one way of passing along information in preliterate societies.

Alternatively, we may say that over time in specific locations, societies had evolved in the school of hard knocks. That is to say, little by little, event by event, they had learned the parameters of risk minimization and the best coping mechanisms for survival. Whether they did this in some Darwinian survival mode or whether they did it through experimentation is often hard to say. On the other hand, we know that many societies in Africa are able to adapt to population pressure by innovation and, indeed, many of the crops they are cultivating were introduced from South America, a practice that represents a substantial innovation. This would tend to suggest that these societies are not locked into mindless ritualistic obedience. Perhaps, then, the tradition represents the repository of experience of a very practical type. Thus, its passage from one generation to another provides the societies with blueprints for survival. In this case tradition becomes a *dependent* variable since it arises from a greater perceived purpose.

As we shall see, the independent variable school held that there was really nothing to be learned from such "static, nonscientific systems", and worked hard to sweep old ideas aside by demonstration, coercion, "education", or other means. Since these traditions were long-established survival mechanisms, this assault on them was unfortunate, and too often we are trying to mine these systems for information only now, at a time when they have been severely dislocated and when much of the knowledge has been distorted or lost. We have to accept, and it is a fundamental premise of this book, that people are rational, and that if they were not, they would not have survived. Of course, there is still the question: "How well adapted are these traditional methods to the new pressures of population growth, commercialization, and expectations?"

Natural Resources as a Determining Factor

It is clear from what we have said earlier that the countries of the tropics are not at the low level of advancement in which we generally find them because of natural resource allocation. If this were the determining factor, aid would be flowing to Japan, not Zaïre. We shall proceed on the understanding that change is possible and that environmental determinism is a convenient blind to the need for better understanding.

REFERENCE

1. Hutton, C., and R. Cohen. "African Peasants and Resistance to Change," Third International Conference of Africanists, Addis Ababa, Ethiopia (1973).

OTHER USEFUL READING

• Brown, L. R., and E. Wolf. "Soil Erosion: Quiet Crisis in the World Economy," Worldwatch Paper #60 (Washington, D.C.: Worldwatch Institute, 1984).
• National Research Council. Ecological Aspects of Development in the Humid Tropics. (Washington, D.C.: National Academy Press, 1982).

CHAPTER 4

Energy and Food

What is agriculture? Quite simply we may look upon agriculture as people intervening in nature to redirect solar energy through preselected plants and animals to produce food. In a way it is a concentration of energy along pathways that are useful to the survival of populations in different areas. It is also important to remember that agriculture involves a manipulation of natural processes. In precapital societies, the manipulation consisted of redirecting energy that is flowing through the system *locally*. In advanced societies, agriculture increasingly involves the *importing* of energy across geographical boundaries of the farm in the form of machinery, fossil fuels, fertilizers, pesticides, and so forth. Energy is the key to the whole process, and by using this parameter we are able to reduce all agricultural activities to a common denominator and then compare them. Since so much of what has passed for "development" in the tropics has involved making their agricultural systems more like those in the West, we shall compare the way energy is used in both types of farming systems. We shall also look at the way energy is used in the tropics and consider the limitations that its use puts on agricultural activities.

Energy Characteristics of Precapital Agriculture

Of course the source of energy that drives all agriculture is that which flows from the sun, and there is little we can do to modify that. We shall not discuss solar energy per se in this chapter, other than to recall that in different parts of the tropics there are differences in cloud cover, solar intensity, etc. These, in turn, afford varying possibilities for crop type.Throughout most of the tropics, and indeed in developed countries before the industrial-agricultural revolution, which started in ernest around 1760, production from the land was largely constrained by the amount of energy that was produced directly by humans and

47

animals. That is to say, it was constrained by human labor, with the occasional use of draft animal power. Most of the labor was directed toward creating the most favorable conditions that enabled nature to work for the farmer. Even though as much as possible was left to the natural processes locally, the farmer had to engage in clearing, planting, weeding, and harvesting. The human labor constraint severely limited the amount of land that could be worked by any individual, which has been previously noted as being approximately 2.5 acres. In precapital times, around 95% of the total energy expended by farm families came from human labor. As we shall see, in modern farming the same figure is supplied by capital inputs.

The traditional farmer has not only to expend human energy for almost all the cultivation processes by which nature is modified, but has to take into account the need to feed those members of society who are unable to feed themselves, such as the very old, the very young, and the disabled. The actual amount of energy required will vary depending on the resource endowment of the region. In the equatorial belt, for instance, it is possible to harvest tree crops such as plantain, which may go on producing with little effort for a considerable time. Thus it was that the farmers of Buganda (that part of Uganda around the northern shore of Lake Victoria) were able to have a fairly easy time of food production and were able to produce a surplus that supported a differentiated society of kings, courtiers, musicians, messengers, and the like. For the rest of the people in regions of seasonal agriculture, more labor was required for land clearance, planting, and harvesting on an annual basis; something close to 1200 hours annually per acre, per person, is a common figure for annual, unirrigated grain systems. We noted that in irrigated systems, it was possible to achieve much higher yields per acre, but at a higher cost. Much *more* labor per ton of crop is required to maintain the water distribution system, to transplant crops from nursery to field, and to dig out drainage ditches.

Before they learned the skills of manipulating the environment by agriculture, people lived by hunting, gathering, and fishing. These systems provided food using minimal human energy since there was no need to expend any effort on changing the natural landscape. On the other hand it was necessary to spend long hours searching for berries or hunting animals, which is one reason these people knew nature so intimately. Where agriculture is practiced, we may observe that populations, in general, rationally opt for the system which, *at their level of technological knowledge, produces the safest yield for the least effort*. Sometimes, usually due to population pressure, this is no longer possible (as we shall see when we look at population in Chapter 7), and then the population has to invest more of its energy per ton of crop into changing and controlling the environment, which then allows them to produce enough food to maintain but not improve their living standards and the security of their community. The greater the pressure, the more energy is required to continue the *same* flow of food. This is like Lewis Carroll's White Queen, who was always urging Alice to run faster in order to stand still. To the eyes of the Western observer, these highly altered terrace and irrigation systems always appear more impressive because of the illusion of "mastery over nature" that they suggest. However, we must not lose sight of the fact that they are far less *efficient* in energy terms than their simpler counterparts. The reason is that the simpler systems put nature to work for the farmer, who is working *with,* rather than *against,* the local environment.

We have seen on numerous occasions in history that people will revert to the simpler, less manipulative system when the pressure is off. This happened in East Africa when, following colonialism, some people along the southern highland areas of Tanzania became less confined following the control of indigenous and Arab slavery, and they reverted to bush fallowing methods after having used intensive green manuring. To put the argument another way, sometimes the traditional farmer had to sacrifice productivity of labor for productivity of land when land became the factor of production in relatively short supply. The intensification represents a decline in living standards in terms of the amount of effort required to maintain the same basic standard of living. So precapital agriculture ties people closely to the land, allows little surplus without a lot of additional human effort, and severely constrains the ability of the individual farms to feed a surplus population.

As mentioned in Chapter 3, one of the only ways to generate a surplus from these systems was by intimidation of some sort using strong control. Sometimes this was effected by domestic or nondomestic slavery where the choice to labor, or not to labor, was removed from the individual. In the case of the Inca Empire, a surplus was specifically grown for the imperial priests and officials as a consequence of powerful obedience to a divine system and pressures that applied both to the present and the afterlife.

Most farmers at an early stage learned that there has to be an equilibrium in the flow of energy through the production system, though they may not have expressed it in exactly those terms. They learned to recognize when more energy was being extracted from the system than the system could sustain over time. Then it was time either to move on, and allow the energy cycle to reestablish itself or, alternatively, to start investing more human labor to replace the energy deficit by, for instance, collecting vegetation from outlying areas to bury or burn on the farm, or by other intensification methods.

Intensification is just another way of increasing the input of energy into the system over a fixed area. Often farmers looked for the intrusion of certain types of grass (*Imperata cylindrica,* for instance, in parts of Asia) or other plants ("index" or "indicator" plants) that were indicative of the onset of energy exhaustion. These were signals to change one's ways or move, and the signals were, wherever possible, faithfully observed. When we come to look at shifting cultivation, we will see a rational adjustment to the realities of the energy cycle, as well as a well-balanced energy management system. In contrast, the colonial powers, and their contemporary successors in government in the tropics, despaired of shifting cultivation because it used so much land and "wastefully" chopped down the forest. Outsiders often missed the point that shifting cultivation incorporated a regenerative period that made it sustainable unless population growth got out of hand.

Table 4.1. Energy Inputs and Outputs in Corn Production in Mexico Using Manpower Only

	Quantity/ha	Kcal/ha
Inputs		
Labor	1,144 hr	589,160
Axe and hoe	16,570 kcal	16,570
Seeds	10.4 kg	36,608
Total		642,339
Outputs		
Corn Yield	1944 kg	6,901,200
Kcal output/input	—	10.74
Protein yield	175 kg	—

Energy Inputs Per Hectare in U.S. Corn Production

	Quantity/ha	Kcal/ha
Inputs		
Labor	12 hr	5,580
Machinery	31 kg	558,000
Diesel	112 L	1,278,368
Nitrogen	128 kg	1,881,600
Phosphorous	72 kg	1,881,600
Potassium	80 kg	128,000
Limestone	100 kg	31,500
Seeds	21 kg	525,000
Irrigation	780,000 kcal	780,000
Insecticides	2 kg	86,910
Herbicides	2 kg	199,810
Drying	426,341 kcal	426,341
Electricity	380,000 kcal	380,000
Transportation	136 kg	34,952
Total		6,532,071
Outputs		
Corn yield	5,394 kg	19,148,700
Kcal output/input	—	2.93
Protein yield	485 kg	—

Reprinted from Pimentel, D. and Pimentel, M., *Food, Energy and Society* (London: Edward Arnold, 1979). With permission.

The message for us from any study of energy is that energy is the ultimate truth. One may be able to do all sorts of things with economics, but where energy is concerned we are dealing with natural processes that are disobeyed at our peril. Any farming system, to have any merit, must be sustainable in energy terms.

Energy Use in Traditional and Modern Farming

Perhaps the best way to start a consideration of this topic is to look at some figures in which the two types of farming are compared in energy terms. Table 4.1 is a microcosm of a whole series of trends that have occurred in agriculture over the last 250 years, and it is well worth our time to consider it in some detail. The crop considered for both systems is the same: corn.

If we look first at the amount of labor per unit of land (the table uses hectares, which are about 2.5 ac), we see that the Mexican farmers have to expend more than 92 times as much energy in terms of labor as their U.S. counterparts, and this involves working in the field under the sun, not in the air-conditioned cab of an agricultural vehicle. This is a phenomenal statistical difference. If we look at the amount of capital involved, we see that for Mexico, it consists of the energy tied up in the production of two simple tools. For the American farmer, it stretches through an enormous array of inputs: machinery, chemicals, fuel, pipes and pumps, and transport. When we compare the energy values of all inputs other than labor, we find that the U.S. system uses 123 times more energy than does the Mexican one. We have to remember that there is energy not only in the items applied directly, such as fertilizers, but in the production of such things as tractors, which are themselves the result of a process consuming huge amounts of energy. Overall the difference in the *total* energy inputs (labor and capital) puts those of the United States at 10 times the figure for Mexico.

When we look at the output side of the table, we see that the yield of corn per hectare in the United States stands about three times higher than that in Mexico, even though 10 times more energy went into producing it. (We can see this when we look at the ratio of energy inputs to outputs). The Mexican farmers are five times more energy efficient than their U.S. counterparts. This is because the system of farming is more intensive in Mexico, since the farmer requires as large a yield as possible from a small area of land, albeit with human labor only. American farmers usually have relatively large holdings and are less concerned with high yields over small areas, though their overall yields are higher.

If the U.S. economy valued all things on the basis of energy, which it patently does not, then the Mexican system becomes more attractive. Conventionally we would identify the problem with the Mexican system as being the **productivity of labor**. That is, the Mexican farmer produces about 1.4 kg (about 3 lb) of corn for each hour worked, while in the U.S. the figure is around 449 kg (about 988 lb) — a ratio of more than 329! This clearly allows the American farmer to produce a massive surplus and feed many people off the farm, which is how they are able to support people in towns and industry. The surplus releases many people from the land for employment in manufacturing and service industries, an occurrence Westerners associate with "development".

A related observation we may make from the table is that the farmer in Mexico is recycling energy largely from within the local system. In the United States, on the other hand, we see a long list of items that do not come from the farm and that have to be purchased, and this in turn requires the system to be commercial and the crop to be commoditized rather than have it feature in a subsistence economy feeding the farmer and the farm family only. The farmer in the United States is, at the same time, locked into a gigantic web of dependence on other people to produce, supply, and maintain the various capital inputs, and this requires a vast expenditure on infrastructure. The farmer also has to know how to use, operate, and maintain the various capital components. If anything external affecting the infrastructure should change, such as shortages or a sudden increase in the cost of an input, then the farmer is at the mercy of the system. The modernization of farming in the tropics, if it is to follow the Western model, will require a tremendous effort to construct this infrastructure efficiently; it is not just a matter of "adding capital".

Further, we should note that the U.S. system incorporates many things that are not part of a natural system. Indeed they are often derived from nonrenewables, incorporate nonrenewables, or are themselves nonrenewable, and this raises questions about the sustainability of the system. The Mexican system is vulnerable to the exhaustion of land through overuse and nonavailability as population grows.

In short, capital allows the farmer to extend the area under cultivation, to raise the yields per unit area, to produce a surplus, and to free up large numbers of people to do other things, while at the same time providing the means to feed them.

The Energy Revolution

We are all familiar with the term "industrial revolution", which provides a label for the events occurring from around the middle of the 18th century. However, it might be more accurate to talk about this period as being an *energy* revolution, of which the industrial revolution was but a part. From the control of steam power onwards, we have seen the application of energy to increasing human productivity in terms of output of goods and agricultural produce and in the transportation of those goods around the globe. If we look at the history of today's industrialized countries over this period we see a gradual transfer of the burden of production from people and animals to machines, fuel, and chemicals. We may consider this to be a general evolutionary pattern, and some writers consider energy consumption per capita to be an index of development (Figure 4.1). So places like Bolivia and Ghana today are on the low end of an energy-use curve along which the developed countries have moved. On the other hand, if the people of the developing world, who constitute more than three quarters of the total population of the earth, were to use energy the way the United States consumes it, then fossil energy resources would be rapidly exhausted. It is worth remembering that the United States, with but 6% of the world population, consumes about 30% of its total energy. Simply copying the energy revolution of the West seems to be a highly speculative and dangerous proposition.

By looking at U.S. farming we can understand the basis of the energy revolution. One good way to do so is to look at gasoline. In late 1991, a gallon of gasoline sold in the United States for around $1.10. Based on a minimum wage of approximately $4 per hour, this means that a gallon of gasoline may be purchased for about 16 min of labor. However, in human energy terms, it

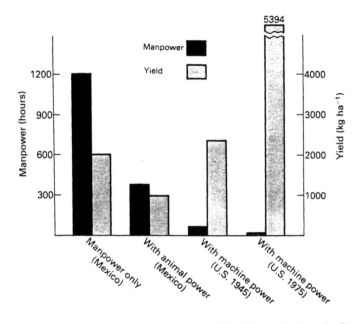

Figure 4.1. The impact of technology over time. (From Pimentel, D. and M. Pimentel. *Food, Energy and Society,* London: Edward Arnold, 1979. With permission.)

is known that a gallon of gasoline is the equal of about 97 hr of labor. One hour of work at minimum wage, therefore, enables workers to purchase about 3.5 gal of gasoline, a quantity that is, in energy equivalents, an astounding 350 times their labor energy equivalent. This is, of course, because of the insane way we price a wasting resource such as gasoline. It is not priced according to its intrinsic energy value, but to a vaguer "economic" value that does not really account for the fact that it will be used up one day.

In the United States, 1000 kcal (about 3 lb) of sweet corn sells for about 15 to 20 times the price of the energy equivalent of gasoline. This gap between the value we put on fossil fuels, compared with the value we put on our daily food, enables us to pour energy into developed farming in this unsustainable way. At present something like 10 units of energy are put in to produce one marginal unit of extra production in U.S. agriculture. All this could change as fossil energy becomes scarcer and more expensive. But we are left with the dilemma of increasing the productivity of tropical agriculture to cope with rising population and rising expectations. Does the Western model make any sense as the basis of transforming the tropics?

Energy Use Around the World

We said, at the outset of this book, that the tropics is not a uniform place either climatically or economically. It is not instructive to construct a ratio of

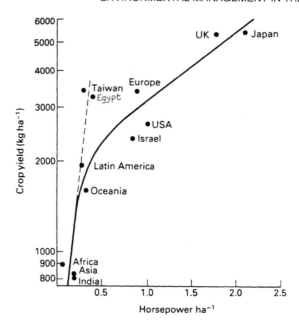

Figure 4.2.   Horsepower and staple crop yields around the World. (From Pimentel, D. and M. Pimentel. *Food, Energy and Society*, London: Edward Arnold, 1979. With permission.)

yield to horsepower because the yield figures are total and do not relate to the specific application of machinery. However, we may make some observations (Figure 4.2). In Asia and Africa, crop yields per unit area are low — one sixth of what they are in the United Kingdom. Interestingly, a huge difference in the application of fossil fuel does not produce the anticipated productivity in Asia, which might suggest that the fuel is not being used effectively or efficiently. Latin America, with about the same level of fuel consumption, produces double the yield. Among the developed countries, the United Kingdom produces the highest yields, half as much again per acre as does the United States, relative to fuel energy unit consumption, but this may be due to the shortage of land and the greater impetus to get higher yields per acre. In the United States, energy prices are low and farms are large, so there is less need for the higher ratio of output per unit of fuel. The case of Taiwan is interesting because of the heavy preponderance of rice, which is high yielding, and the scaling down of machinery to suit the smaller farms. Looking at the curve, one gets the overriding impression that modernization has been associated with the harnessing of more horsepower per unit area, resulting in higher yields, with the interesting exception of Taiwan.

By reducing everything to the common denominator of energy, we may gain a better insight into the components of agriculture and the comparison of agricultural systems. We have seen that though the traditional farm is labor efficient in the sense of output/input, it is ultimately constrained by its low level of labor productivity. This will become critical as population numbers increase. We have also seen that farmers may invest labor into capital formation, but this reduces the efficiency of labor — though without the alternative of Western energy-intensive capital, the farmer may have no choice. We have also seen that the quantum leap in agriculture came when large amounts of energy were brought into the system to supplement the productivity of labor.

USEFUL READING

- Pimentel, D. and Pimentel. M., *Food, Energy and Society* (London: Edward Arnold, 1979).

CHAPTER 5

Some Principles of Traditional Land Use

We noted in Chapter 3 that traditional farmers often survived by mimicking the processes that could be observed functioning in the natural ecosystem around them. In addition we need to understand some of the basic influences that shaped the activities of people using the land for their own livelihood. These principles have a universality and relate to the practices prevailing before the introduction of massive amounts of energy across the frontiers of these land-use systems. In the context of these ideas, the activities that we see have a profound underlying rationality, but they were not always understood or appreciated by those who came to the tropics with the intention of turning these areas to commercial advantage on an international scale.

The Short Time Horizon

The tropical environment, with its high temperatures and its active disease and pest conditions, is not conducive to the long-term storage of crops. Family farms often feature a small granary, or store, to house the crop that will keep the family through the hungry season in the dry tropics and provide the seed for the next harvest. However, the conditions generally do not favor the retention of crops much beyond the coming year. Provision for a period beyond that time has to rely not on storage based on above-ground crops, but in — or under — the ground crops. An example would be the cultivation of a drought-resistant variety such as cassava (*manioc*), which has little food value beyond starch, but which can keep a family alive *in extremis*. Furthermore, the production of large, sustained surpluses is extremely difficult within the constraint of human labor, as we have seen from Chapter 4. The farmer is concerned, therefore, in charting a course of survival through the coming season, with some emergency provision of a nonstored nature beyond that.

The Nature of Risk

Except for those people living in the humid tropics who are able to derive their subsistence from perennial crops, the concept of risk is a very *immediate* one. In the upcoming season, the farmer has to consider shortage, or lateness, of the rains; the possibility of disease affecting the crop or the herd; ravages by animals; declining fertility of the soil with continued use; damage by weather; and theft. In the absence of large quantities of stored crops, the production system itself has to provide some sort of insurance cover against these difficulties. Over time farmers have devised elaborate strategies or *coping mechanisms* to deal with these risks, and we shall examine some of these later in this and in following chapters.

These coping mechanisms become enshrined in the traditional practices of the community, including their often ritualized adherence to certain crops, the community, and methods of cultivation. It is worth noting here that these methods have ensured the survival of the community over time, and so there is a tendency to be protective of them and to identify with them. This makes these communities appear conservative, or reluctant to change, as many a colonial adviser discovered. However, they are not completely opposed to change, and we have observed that they have often, in relatively recent times, changed their basic crop where some other crop offered a distinct advantage.

It is obvious, though, that where the survival of the family or the community is concerned, and where the risk factor is high and immediate, there will be a natural caution about change unless there is a high degree of certainty that the change represents a real improvement over what existed before. It is often stated that there are leveling mechanisms or barriers to change in such societies. For instance in Melanesia, farmers owning large numbers of pigs, which makes them conspicuously "wealthy", are expected to throw lavish parties that reduce the number and redistribute the wealth; in Fiji, property is held by the village, not the individual, making the Western model of personal advancement difficult. In Zambia, families accumulating large numbers of cattle are expected to slaughter many of them when the occasion for a funeral arises. In the Zambian case, the funeral also allows people keeping cattle to share the meat, since an entire carcass is too much for the average family to handle. These mechanisms emphasize the need for identity with the community, which is itself a form of insurance, and with the practices that have traditionally supported that community's survival. Change should be seen in this context.

Minimum Risk for Minimum Labor

No farmer anywhere in the world is interested in working more than is necessary to achieve the desired output. Tropical farmers are the same as all others. However, they will generally labor as much as is necessary to minimize the risk to themselves and their families. At this point it is useful to note that, again with the exception of the perennial areas, no farmer or animal herder can work on the basis of "average conditions". As we observed earlier, the drier it gets, the more uncertain the chances of receiving "average" rainfall become. Principally, cultivators must work on the basis of getting through the *worst* conditions to the best of their abilities, and within the technological limits of that society. This is why many farms and herds will contain an *insurance* factor. To the outsider this may appear to be a "surplus" since it may be in addition to the requirements of the family when conditions are favorable, especially if that is the situation when the observer happens to drop by. This was often the way that colonial administrators viewed the cattle herds of African nomads. However, those communities often needed every animal, or every head of grain, for survival when times were bad. Any appraisal of a subsistence farm or herd must allow for this risk factor.

The "Poor But Efficient" Hypothesis

The American anthropologist Charles Schultz, writing about the peasant farmers of Central America, accurately described them as "poor but efficient".[1] By this term he meant that in the context of their precapital technology (what he called the "state of the art"), they combined labor and land in such a way that it is almost impossible to achieve their aims more *efficiently* by any recombination of these factors. As long as one accepts the innate rationality of people, this hypothesis may seem rather obvious. However, it did not figure very prominently in the minds of those who wanted to develop and change traditional farming, who so often perceived these farmers as perverse, lazy, reactionary, and the victims of mindless ritual. Indeed when we look at these communities through the eyes of travel magazine writers, the focus is often on the bizarre or ritualistic elements of the communities.

Many anthropological studies, right up to the 1960s, were written about communities in a way that left the reader wondering how the people in these books stayed alive, though we usually knew everything there was to know about their kinship structures. The implication of the Schultzian hypothesis is that it is well-nigh impossible to change the relationships between the factors of production to produce an increase in efficiency (output/input) unless one introduces some capital element. A good example of this was provided in Uganda some years ago when the agriculture department emphasized the 30% gain in cotton production that could be achieved if only the farmers would plant the crop earlier, as they had been advised. This analysis disregarded the fact that "earlier" was the time when they were planting their millet and that there was no spare labor capacity available then. Cotton was something that the farmer had to grow to pay taxes. Millet was the crop that kept the family alive, and it was obvious where the farmers would place their emphasis.

Coping Devices

We shall look at many coping devices as we look at different examples of land-use systems, but it is useful to summarize some of them at this point. Rarely did a community rely on just one of these; instead these mechanisms formed an interlocking web of insurance in a situation where there were no international relief agencies, often no real central government able to do anything, poor communications, and no banks. Sometimes it was difficult for the outsiders who generally wrote about these systems to see the coping mechanisms for what they were. They were seen as ceremonial instead. But the reverse could be said about Western traditions. An outsider might wonder about the function of an American wedding where people dress up in ridiculous clothes, parade around, stand up, sit down, sing, and so forth. For the participants, it has a very real purpose: to emphasize the social bonding that holds (or held?) our society together, provides a support network for the upcoming generation, and binds the members to the religious and social community with which it identifies. It is more than just the "big day".

The Social Network

Western society is used to the idea of separating *social* and *economic* elements into two, distinctly different groups. Indeed we teach these as different disciplines in schools and universities. In a society where so many transactions are made through the medium of money, perhaps this is appropriate. What is most manifest in a traditional community is the personal nature of most relations, as compared with the essentially impersonal nature of transactions in our society. In precapital societies, it is much more difficult to make this distinction between social and economic roles. Looking at the organization of the extended family in Moala, Fiji, Marshall Sahlins described the relationship between the way groups of people live together and the way this grouping relates to the tasks hand:

> The fundamental activities of the extended family are economic: the members form a labor pool; property and produce are pooled in providing for the common hearth; and the internal ranking scheme is primarily a means of organizing production and distribution. The women . . . care for the children, keep the houses in order, prepare meals (which sometimes means gathering firewood and vegetable greens), make mats, do most of the fishing, and collect seafish, sea slugs. . . . The women are organized as a cooperative labor unit. The men also form a labor pool. The men's primary tasks are gardening, housebuilding, and some fishing. In earlier times, the men of an extended family often formed a work unit for house-building, clearing land, and firing it, digging irrigated taro patches, and planting and weeding gardens. . . .
>
> The size of the [extended] family made it possible to release some members for work on distant fields without hardship to those left in the village. A common cookhouse and common meals and centralized supervision of the gardens ensure that the different foods will be shared among all members regardless of their particular contribution to production. . . . Cultivation of the distant gardens is undertaken by the youngest and strongest. . . . This type of family group is an ideal unit for working scattered resources without sacrificing any of the usual familial functions of child care, socialization, and the production of capable, mature members of the society.[2]

Elsewhere, Sahlins notes:

Economy becomes a category of culture rather than behavior, in a class with politics or religion rather than rationality or prudence: not the need-serving activities of individuals, but the material life process of society.[3]

Consequently, changes in production methods may well require parallel adjustments in the social organization, as with the tendency toward individualization that comes with the conventional model of modernization. Put another way, in traditional societies, though some products may be traded (see later), factors of production (land and labor) are almost never put into the marketplace.

Groups in the community will come together to perform tasks on preset dates to the accompaniment of specific songs, dances, and beer drinking. The whole thing has the appearance of a ceremony, but it may be the way that economies of scale are achieved in clearing land, building homes, or digging irrigation ditches. It is the social duty of members of the community to do this, much in the way that the Amish will still put up a barn in a day. In general each family has to be able to carry out nearly all the processes of agricultural production from beginning to end, but there are occasions when the whole community may be more than the sum of its parts.

In most societies it is instructive to look at the way that kinship (the formalized relationship of one member of a distinct community to another) is organized and how it functions. People in the West have a well-defined kinship structure, though its meaning has changed quite a lot over the decades. To a large extent it is a support mechanism since there are members of any society who may be unable, permanently or temporarily, to support themselves, such as the very old, the very young, the sick, and so forth. Western society has increasingly passed that responsibility over to the fabric of the state, though many people in a traditional context would find that incomprehensible, as would our ancestors, no doubt. This abrogation is not an option in most traditional societies, and the community has to take care of its own. This is one reason often put forward for the vesting of all property in the community in Fiji and other places. The community, and one's identification with it, is a form of insurance and that is one reason why people may be reluctant to stand out from the crowd and why the colonial administrators despaired of finding "progressive farmers" among such people.

The kinship structure represents, and is, an interlocking web of responsibilities and obligations that helps carry individual members through difficult times: crops may be shared; animals loaned out; children adopted; scale economies achieved; and communities defended in times of war. How kinship is structured may depend on the circumstances of production, so it is unlikely that the same system will be found among nomadic groups of pastoralists as will be found among settled farmers. But without this network the individual really has nothing, and that is why the community may be prepared to pay a high price to support its priests, chiefs, and others who provide the symbolism of identity and embody the ritual that provides for continuity. That also is why traditional communities often seem to constrain individualism, since it threatens the communal basis of survival.

Diversity

When we come to look at one of the simplest production systems, shifting cultivation, we are struck by an appearance of seeming chaos. This is not the orderly monoculture of Western farming. On some small farms, as many as 90 crops may be seen under cultivation. On the other hand, everything has a purpose and a place. By maintaining this diversity, the farmer is diffusing the risk of disease, drought, and predation over a broad spectrum, and so is unlikely to face ruin if any one of these difficulties arises. The range of crops also provides a diversity of inputs for the diet, produces fuel and building materials, combats drought, and may provide food for the animals. Furthermore, it exploits the range of nutrients available to the fullest degree and may adjust to declining fertility as the cultivation proceeds, prior to the abandonment of the plot.

Another form of diversity is topographical. Quite often farmers will have pieces of land in different locations. This method was extremely well-developed along the slopes of the Andes, exploiting the different ecological possibilities that came with altitude (crops, grazing for the *alpaca*). Forty different species of crops were adjusted to different altitudinal ecological niches by the Inca communities in these areas.[4] Again this was often perceived as "land fragmentation" by the colonial powers, who saw it as a result of inheritance laws and who spent much time calculating the time that farmers "wasted" walking from one plot to another.

Indeed, such a topographical spacing allows farmers to exploit different soil types, vegetation types, aspect of the land, proximity to water, and a range of other factors. In hilly areas, for example, it was common for the farmer to have a piece of land on the hilltop as a woodlot and for occasional grazing. Below that, pieces of land would be arranged along the different configurations of soil adjusted to the slope (the *catena*). Around the house would be an intensively cultivated garden providing many of the daily needs and often incorporating the household waste as a supplement to natural fertility (Figure 5.1). At the bottom of the slope would be another piece of land for drought-resistant crops, watering livestock, or for possible irrigation during a dry season. The river itself provided fish and, possibly, a medium of communication. Maintaining a mix of livestock and crops was another aspect of diversity. Even within plots, often minute adjustments to soil and water conditions would be made to favor those crops that did best in a particular spot. This is illustrated also in the concentric way that the use of land would be organized around the household, as shown in Figure 5.2.

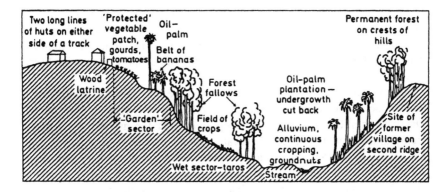

Figure 5.1.   Diagrammatic cross-section of Pene Niolo, Congo-Kinshasa, and its vicinity (1949). (Reprinted from Ruthenberg, H. *Farming Systems in the Tropics,* 1971. By permission of Oxford University Press.)

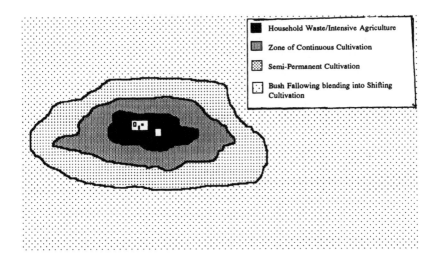

Figure 5.2.   Stylized distribution of land around a settlement in lowland African savannas.

Mobility

In many but not all traditional systems, mobility plays an important role. Of course, the modern Western farmer with a car is particularly mobile, but the farming activity is usually very place-specific because of the huge amount of capital invested in that spot. Before farming began, and still among some communities today (e.g., the Kalahari Bushmen), people survived by hunting animals or collecting naturally occurring foods. But early farming systems relied on mobility, too. To counter the decline in natural fertility of the soil, farmers "moved the fields" to another spot. Nomads, with their animals, moved to water and grazing points as the seasons changed. This mobility helped to overcome the constraints of the natural environment and produced, within the population limits of each system, a sustainable life-support mechanism based on renewable energy.

As population increases, and in general as the pressures for incorporation into "modern" society are imposed, the mobility factor is severely constrained. The drawing of boundaries by the European powers, for instance, was a major factor limiting mobility as people became beholden to a new state power. This put severe limitations on nomadic people, whose production system often consisted of several distinctly different ecological niches in widely separated areas. The system was rarely sustainable in any *one* of these.

In its most extreme form, mobility was demonstrated where population pressure grew too great for the local environment to sustain it without deterioration. In such cases whole communities moved, and this was more common than we may realize. It is thought that the explosion of the Golden Horde (the Mongols) right across Asia and central Europe, especially in the 13th century, was in response to too much pressure on the home area. Sometimes communities, such as the Iban in Borneo, would take up migratory warfare rather than face the pressures of increasing cultivation. Africa, for instance, is a patchwork of major movements of whole communities (as, indeed, is the early history of Europe). And so migration provided a type of ultimate safety valve to relieve the pressures on the environment.

An interesting case of mobility is that of the Inca-period peasants. Though they may be imagined to be closely tied to their terraced fields, in fact many of the communities had outliers in far-distant lands. The people of Titicaca had outliers in the valleys of southern Peru and in the interior of Bolivia at Cochabamba. In these locations it was possible to grow different crops such as cotton, coca, and maize, which would not flourish on the high plateau. The areas were worked by teams that left the plateau seasonally to cultivate the colonies.[4]

Land Tenure

In many parts of the world, the ownership of land was a concept largely unknown to traditional farmers. Nevertheless, the security over the land as a means of ensuring survival was often vested in the community, which alone embodied territorial rights. The individual family usually had specific rights to the use of a piece of land as long as the family members behaved in a way that demonstrated their continuing identity with the larger group. Usually, continued wise use of the land ensured that the family's rights would remain protected. However, with modernization, and the requirement to invest capital in a particular piece of land, the pressures for some form of individualized recognition increased, and were sometimes encouraged by the colonial authorities, as in Kenya. Decisions regarding changes in land use would, to a large extent, be vested in the community and often required consensus or the approval of elders.

Innovation

It is easy to overlook the fact that societies under pressure, or facing unendurable risk factors, also showed the capacity to change their ways.

> Tropical farmers are experienced in adapting their systems and methods of production to different circumstances, albeit slowly, and with only a limited success in making major technical discoveries. . . . Innovation and change, however slow, are normal features of traditional farming. Without a continuous series of small adjustments, these diverse and often well-adapted farming systems could never have been.[5]

For instance, a community that is running out of land can migrate or it can change its relationship with the land. Of course this will be limited by Schultz's "state of the art", but change was often possible. The community could, for example, expand the cultivable area by terracing hillsides, making fields upon slopes that, otherwise, would have been eroded away if cleared. In this way the truly spectacular landscapes of Luzon (Philippines), Nepal, and the Yemen were created. Labor was invested as capital to sustain production over time, and whole landscapes were transformed to a degree almost incomprehensible to us now. In effect it often amounted to sculpting an entire mountain range with hand tools four millennia ago!

Environmental Knowledge or Ethnoscience

Although we have stressed risk aversion, societies practicing early simple forms of agriculture did engage in experimentation. In the Philippines, for instance, the area around the household was used to try out new and potentially interesting varieties of food plants. Usually, however, the farmer relied on a remarkable knowledge of the local environment and its capabilities. Of course the relationship with the environment was often a direct and immediate one, and tradition handed down the wisdom of earlier generations. Often the taxonomies of plants invented by societies locally were more profound than those derived by the Europeans who came later. But overall, survival depended on an intimate knowledge of the world in the immediate vicinity or, in the case of the nomads, several different environments spread over a wide area. In one of the first serious studies of traditional farming, William Allan described the situation concisely:

> The shifting cultivator has an understanding of his environment suited to his needs. He can rate the fertility of a piece of land and its suitability for one or other of his crops by the vegetation which covers it, and by the physical characteristics of the soil; he can assess the *staying power* of a soil, the number of seasons it can

be cropped with satisfactory results, and the number of seasons it must be rested before such results can be obtained again. His indicator of initial fertility is the climax vegetation, and his index of returning fertility is the succession of vegetative phases that follow cultivation. In many cases his knowledge is precise and remarkably complete.[6]

Trade

It is worthwhile spending a few moments to look at the question of trade. Because we have placed such emphasis so far in this book on the large labor requirements needed to offset risk and ensure subsistence, we have not really portrayed a system in which there is much trade or predictable surplus for that matter. When we think of these traditional farming systems we generally imagine whole communities of farmers growing much the same thing and without a well-developed communications network. There is, therefore, not much of a surplus to provide for specialization of labor, though in a good year Mexican peasants sometimes harvested 20 to 100% additional to their needs. Even where there was limited trade, there was often an exchange mechanism, sometimes conducted through the processes of social obligation, which ensured some element of redistribution. Yet there was often more trade than we imagine. If the farmer is growing crops in a seasonal regime, then there is down time during the year for other activities such as weaving, mat making, beer making, and the like. In Guatemala and Mexico, the Indians live in communities distinguished by an economic specialty in addition to agriculture, such as pot making, blanket weaving, or lumbering. Also, in good years there may well be a substantial windfall surplus that may be traded.

The most obvious opportunities for trade exist across major ecological boundaries, such as those where highlands impinged on lowlands, as in Ethiopia and the Yemen or in Central America. Here the different crops could be exchanged as they are today at markets along the break of slope. Another case would be where dwellers along a river could exchange dried fish for produce. The hunters and gatherers of the forests of Zaïre trade their goods with neighboring farmers, often without ever meeting ("silent trade"). In some cases highly mobile communities, such as the Bedouin of the Arabian deserts, exploited their mobility to carry goods from one ecoregion to another. Thus they traded the frankincense of South Arabia in the markets of the Levant. The Shan of Burma traded in this manner, and such markets were common throughout Asia.

In Africa amazing trade networks developed *across* the vast expanse of the Sahara Desert, and great emporia developed such as Timboctou in modern-day Mali, where the products of the Maghreb, the Sahel (salt), and the West African coastal kingdoms were exchanged.[6] In a rather special form of trade, the nomads of the Sahel would move south during the dry season and graze their animals on the stubble of the grain fields; the animals, in turn, provided manure for the fields. One interesting aspect of African trade concerns the metalworkers. These were often specialized communities who jealously protected their secrets and surrounded their activities with mystique. However, they produced the iron that transformed African farming by replacing the digging stick with the hoe, and before 1500, every African hoe was made on that continent. Subsequently, this indigenous aspect of production and trade was displaced by cheap, though superior, European imports.

The trade network provides an interesting insight into the concept of mobility. As we have said, communities in a given area had a limited basis for trade because they tended to produce the same things for their own needs. Nevertheless, throughout most of the tropics, networks of local markets developed, even if only a clearing in a prearranged spot, which functioned as a market on a known day each week or month. Or each day a market might function in a different place on a rotating basis, as was the case in Nigeria. Each market could draw its goods from an area up to a half-day's journey away, and by going to different locations, the producers might gain access to a considerable number of people interested in what they have to sell.

The markets also functioned as locations where itinerant traders could come occasionally, bringing exotic goods from a much wider area. A good example of this would be the Hausa peoples of Nigeria, who traveled over huge distances from market to market trading goods and buying the specialties of each place in which they found themselves. Certain locations might be visited only once a year on these huge circuits, and they became the sites of great festivals. This is, of course, exactly the same process by which medieval Europe distributed its goods through markets and fairs.

In general, in a surprisingly large number of places around the world, markets had a prescribed day, and each town could become a market by charter only if it did not compete with the catchment area of another. On predetermined — usually religious — days, there would be a fair, though this might last a week. In this way specialized communities of traders and artisans could be supported from a meager base, simply because of the enormous catchment areas these mobile systems could encompass. Since traders and artisans had to be fed, there existed the basis for a barter market for local food produce as well as products from the hunt, the home, and the field. In this way was Tanzania able to trade with China in the 12th century, as evidenced by pottery and coins found in Tanzania today. On the other hand, the production systems of these areas were almost never *primarily* geared toward serving the market. The market provided an occasional opportunity to supplement the range of household goods. And, apart from the artisans and traders, few people relied on the market for basic subsistence. Where populations were settled and relatively dense on the ground, as in irrigated areas, then trade was generally more developed.

REFERENCES

1. Schultz, T. W. *Transforming Traditional Agriculture* (New Haven, CT: Yale University Press, 1964).
2. Sahlins, M. D. "Land Use and the Extended Family in Moala, Fiji," in *Environment and Cultural Behavior,* A. Vayda, Ed. (Garden City, NY: Natural History Press, 1969), pp. 395-416.
3. Sahlins, M. D. *Stone Age Economics* (Chicago: Aldine-Atherton, 1972).
4. Favre, H. *Les Incas* (Paris: P.U.F., 1975).
5. Ruthenberg, H. *Farming Systems in the Tropics* (Oxford: Oxford University Press, 1971).
6. Allan, W. *The African Husbandman* (Edinburgh: Oliver and Boyd, 1965).

OTHER USEFUL READING

- Boster, J. "A Comparison of the Diversity of Jivaroan Gardens with that of the Tropical Forests," *Hum. Ecol.* 11:47-68 (1983).
- Bovill, E. W. *The Golden Trade of the Moors* (New York: Oxford University Press, 1958).
- Eden, M. J. and A. Andrade. "Ecological Aspects of Swidden Agriculture Among the Andoke and Witoto Indians of the Colombian Amazon," *Human Ecology.* 15:339-359 (1987).
- Klee, G. A., Ed. *World Systems of Traditional Resource Management* (London: V. H. Winston and Sons, 1980).
- Moran, E. F. *Human Adaptability: An Introduction to Ecological Anthropology.* (Boulder, CO: Westview Press, 1982).
- National Academy of Sciences. *Ecological Aspects of Development in the Humid Tropics* (Washington, D.C.: National Academy of Sciences, 1982).

CHAPTER 6

Adaptations to the Environment

As we look back through history, we see an evolutionary pattern to the way people have used the land. In earliest times subsistence was derived from hunting and gathering, which involved collecting from nature with almost no modification of the natural landscape. Later, around 10,000 years ago, people discovered ways in which they could rearrange natural assemblages of plants to favor those plants that featured in the diet of the community. Initially, agriculture involved a harnessing of local energy processes, though some of the plants might have been brought in from outside the immediate area of cultivation. At its most basic, this early form of agriculture involved running down the local store of energy and then leaving it to regenerate by means of long fallows. Eventually, mobility diminishes as population grows, and the fertility of the area is supplemented by activities involving increasing investments of human labor, in a more settled form of cultivation. Around the middle of the 19th century, dramatic changes began to occur through the application of fossil fuels, machines, and chemicals. At this point the direct link with human labor was broken and productivity increased phenomenally, as noted in Chapter 4.

Although this picture is presented as an evolutionary one, elements of each stage may be seen in different parts of the world today, and the tropics still relies heavily on the earlier stages. However, the main drawback with the traditional systems is that they quickly fall victim to increasing population density, reaching a point at which, in energy terms, they are no longer sustainable in the long run. Before dismissing these as anachronisms, it is instructive to examine the rationales of the traditional systems since they normally incorporate sound principles of land management and because they contradict the often-stated derogatory comments passed by early foreign administrators and agriculturalists. Indeed, many of the principles displayed by these labor-intensive tropical systems are again finding favor in a world increasingly concerned about a dependence on nonrenewable fossil fuel resources and the need for sustainability.

In this section attention will focus on those systems that were adapted to the local characteristics of the ecosystem, lived within its limits, and copied its characteristics. The West has inherited the Judeo-Christian tradition of *mastery over nature,* in the context of which it has extolled the element of control. This battle mentality is still demonstrated in such terms as the *conquest* of space, which is a patently ridiculous idea. We barely know what space is, let alone how to conquer it. Other societies and religions have often emphasized the unity of all things — Buddhism, for instance, or the Native American culture. We must therefore appreciate the strong cultural bias of much of what has been written about tropical agriculture by outsiders, and indeed the bias present in our evaluation of our own systems of cultivation. The real question for us all now is whether or not **any** philosophy can prevent our being overwhelmed by the size and speed of population growth.

Before looking at some representative examples of adaptation to the environment, it will be useful to note the implications of adopting an agricultural system. Once a society moves from simple hunting and gathering, and decides to plant crops or herd animals, it has to face up to a number of ecological realities. By cropping the land the farmer extracts nutrients that are not being replaced in the way they were by the original, more diverse, vegetation. Thus these either have to be replaced by the farmer somehow, or the farmer must face the fact that the fertility of the plot will decline fairly rapidly as the seasons go by. For the system to be sustainable, there has to be an answer to this dilemma. Furthermore, since the farm family is investing its time and effort in modifying, albeit temporarily, the natural landscape, there needs to be some form of protection of the right to use a *particular* plot of land. This does not imply ownership, just user rights. In addition, the family must have a powerful working knowledge of what grows where and how the family may best survive using a select group of plants, rather than the full array that nature would have grow there. This intimate knowledge may not be immediately evident to the outsider, but it is there.

Hunting and Gathering

The hunting and gathering system relied entirely on the bounty of nature in a totally unaltered form. Today we tend to associate this method with extremely remote or harsh environments such as the rain forest of Zaïre, the Kalahari Desert of Botswana, or the interior forests of Indonesia. It was not always so, and for the first 2 million years or so that our ancestors lived on the earth, this was their means of subsistence. To be successful the population must be extremely mobile, and in each area they may find only a few items that are useful for consumption. The system is therefore *extensive* and requires an enormous amount of land to support a tiny number of people. It has been estimated that if hunting and gathering were the sole food garnering technology known to us, the earth would support a maximum population of around 10 million. Following the discovery of agriculture and the domestication of animals 10,000 years ago (the Neolithic revolution), hunting and gathering was steadily displaced from cultivable land, which is why it tends to remain in those inaccessible or uncultivable areas today. In fact, even in the Kalahari, the hunting and gathering method is well able to provide for the food needs of the population with relatively little effort, so that one able-bodied male may provide for five other family members. Nevertheless, such an extremely mobile system does not provide for the development of settled communities, much division of labor, or the growth of marketed goods. Unfortunately this "simple" lifestyle has been viewed as primitive, in the most pejorative sense of that word, by its settled neighbors who have sought to either exterminate these communities or settle them with almost no understanding of how they function. In short, these roving people are seen as an embarrassment and are often, as in Zaïre, treated as a joke. What we now know is that, of all users of the earth, these communities have perhaps the most intensely detailed knowledge of their environment. This is not surprising since they would not have lasted long without it. Only now is any effort being made to recognize this skill and to formalize the knowledge. It seems unlikely that this most ancient of systems will be able to persist for much longer.

Shifting Cultivation

This ancient form of cultivation provides us, if we are willing to put aside our prejudices for a moment, with a fascinating insight into the functioning of precapital agriculture. Shifting cultivation is still the principal means of support for over 200 million people covering 14 million square miles, mainly in the upland forests of Asia (the lowland areas being largely irrigated) and in the great rainforest basins of Africa and Latin America. As with hunting and gathering, this system of cultivation attracted the ire of many a colonial administrator or Western-trained agriculturalist. Typically it was described as wasteful and destructive,[1] but under the right conditions it is neither of these things. The problem is that, to the Western eye, this practice appears to be chaotic and requires the wholesale burning and chopping down of forest resources.

The shifting cultivator moves into an area of forest (the drier equivalent of this system is called bush fallowing), clears away 1 or 2 acres of trees, and often burns over the residue. There is no "field", in the typical Western sense. Instead the plot contains some uncleared trees, remains of the trees not wholly burned, termite mounds, etc. In this cleared area the cultivator plants a variety of different crops. After the first season, the yield of the first round of crops will start to decline because, essentially, the sparse fertility of the forest is being mined and there is no canopy of trees there to replace it. The initial burning will have released some of the energy contained in the system and will have supplemented, for instance, the potash in the soil through the ash. The farmer makes an effort not to leave the ground bare and exposed to the elements because that would lead to leaching and an even more rapid decline in the fertility. During the second and third seasons, the cultivator may change the combination of plants in recognition of the soil's diminishing fertility. Eventually it will become evident, usually through the incursion of certain "indicator" plants, that it is not practical to continue in this location.

At this point the entire farm shifts and the process starts all over again. Meanwhile a slow process of natural regeneration takes place, encouraged by the fact that the original plot is still surrounded by the forest vegetation. Eventually something closely resembling the initial vegetation cover will reestablish itself and the fertility level of the thin humus layer will be restored. At that point it will be possible to use the plot again. Once more, the cultivator is able to recognize, from the plants that reappear, whether this regeneration has run the full cycle.

So, for any farm family we are able to construct a simple calculation. Assuming that the family is able to cultivate a plot for 3 years and that it takes 15 years for the fallow cycle to run to completion, and assuming the farm occupies about 2 acres then we may say that for the individual farm the total cycle takes 18 years. Of course this will vary depending on the quality of the soil. We may also express this relationship a few other ways. For example, the number of acres that each family requires to farm without permanently damaging the environment is the number of acres that constitute the typical farm multiplied by family size (approximately six persons) and multiplied by the length of the cycle (the number of years required for regenerating the level of fertility during fallow). We then divide this first figure by the number of years of cultivation for each plot. In the case outlined earlier, this means that the family requires 12 acres (though the figure may be much higher with poorer soils). From this calculation we may also state that the maximum number of people this system will support per square mile (640 acres) is 640 divided by 12, multiplied by the size of the family. The result is 320 people. Beyond that point the system will break down. Clearly, this is a system that requires a great deal of land per family, which may be why it was described as wasteful by some writers. On the other hand, below this level it is not wasteful since it is recycling local energy resources using natural processes.

The organization of the farm, or more strictly the garden (since the word "farm" implies something much more "controlled"), is extremely interesting. At first glance it appears to be a picture of chaos. There is no monoculture, which is what the West often associates with farming. There appears to be an enormous jumble of different plants scattered all over the plot. Closer examination reveals a different picture. The farmer exploits every niche in the small environment the family inhabits. This is to take advantage of the microvarieties of soil type, fertility, etc. On some farms as many as 90 different varieties of planted crops have been counted. Such variety provides the family with a range of foods including grains, legumes, vegetables, and flavorings. Writing about the Hunanóo of the Philippines, Chester Conklin wrote:

> A study of Hunanóo soil classification and associated ideas regarding suitability for various crops . . . checked well with the results of a chemical analysis of soil samples. Ten basic, and thirty derivative soil and mineral categories are distinguished by the Hunanóo farmer.[2]

In addition, the variety of plants provides a hedge against risk, as exemplified by the starchy tubers that are often planted in case a significant part of the crop fails or is damaged by wild animals. It also allows the farmer to spread the work through the year and similarly to spread the harvest so that there is always something for the family to eat. The remarkable range, diversity, and sequencing of crops on a small farm in Mozambique is shown in Table 6.1.

Another aspect of the shifting cultivation system is its adjustment to declining fertility. It is not possible for the family to maintain the same range of crops over several seasons. A first crop of maize may exploit the initial reservoir of fertility, but it will be so demanding of nitrogen that it may not be continued. The adjustment to declining fertility is shown in Figure 6.1. Yet another adaptation is in the **interplanting** of crops so that it may be possible for one to reinforce the other (shade-loving species for instance). Around the hut will be a vegetable garden, often using the domestic waste produced by the family.

The attraction of this system, apart from the fact that it puts nature to use without long-term depletion of resources, is that it provides a high return to labor. After the initial clearing and planting, there is a small but steady demand on the family's time. It has been estimated that about 30% less time is required, for the same return, for shifting cultivation than is required for its settled counterpart (without capital inputs).[3]

Within its population constraints this system flourishes, but the mobility of the people and the farms, along with the necessary low density of population, means that it is difficult to set up a modern marketing system or to incorporate shifting cultivation into the commercial economy. Governments, therefore, have a tendency to perceive this system as not playing its part in development. As will be noted later, governments seem to have a natural aversion to any group of people that is constantly on the move; for one thing, they are difficult to tax! (Consider, for a moment, the European attitude toward the Gypsies.) There is an additional concern about a system that relies so heavily on burning as a tool of farming. Fire was one of the reasons why the administrators looked upon shifting cultivation as destructive. Where the population limits are exceeded, there is no doubt that shifting cultivation does become destructive, but it is not inherently so. Overall, shifting cultivation has received a negative response from those who have written about it since it is so difficult to incorporate within the modern state economy. It is true that, unlike its irrigated counterpart, shifting cultivation is unlikely to lead to a settled civilization of towns and artisans. For many years writers have argued about the Mayan civilization of Central America, which used a form of shifting cultivation known as *milpa*. It has been suggested that this system broke down under pressure of weight of population, which is why the Mayan civilization disappeared. This is far from proven, however. The main attraction of the system remains the fact that it is possible to obtain 10 to 20 times the amount of food energy in return for the labor inputs.

Table 6.1. Consumption Chart of a Shifting Cultivator's Family in Manhaua, Mozambique

Product	|	Month 1	2	3	4	5	6	7	8	9	10	11	12
Crops produced by the family:													
(1) Staple foods containing starch													
Manioc		×	×							×	×	×	×
Maize in milk-ripeness				×	×								
Maize as ripe corn						×	×						
Rice							×	×					
Sweet potatoes							×	×	×				
Sorghum									×				
Sorghum-corn (ecununga)									×				
Sorghum-cane (maele)							×	×					
(2) Staple foods containing protein													
Beans (boer boer)										×	×		
Beans (jugo)							×						
Beans (manteiga)							×						
Green beans (boer boer)								×	×				
Green beans (nyemba)					×								
Green beans (jugo)					×	×							
Manioc leaves		×	×	×	×	×	×	×	×	×	×	×	×
Sweet potato leaves							×	×					
Bean leaves of all kinds			×	×	×	×	×						
(3) Additional foods and spices													
Onions												×	×
Tomatoes						×	×						
Gherkins				×									
Eggplant (two kinds)											×	×	×
Quiabo (*Hibiscus esculentus*)			×										
Groundnuts							×	×	×	×	×	×	
Sugar-cane				×									
Pumpkins		×	×	×									
Sorghum-cane (ecununga)						×	×						
Collected food:													
(4) Staple foods (animal protein)													
Grasshoppers		×	×	×						×	×		
Mice and other rodents							×	×	×				
Caterpillars					×	×							
Larvae of butterflies												×	×
(5) Additional foods													
Fungi (nine kinds)		×		×									×
Wild fruits (seven kinds)		×										×	×
Honey						×	×	×					

Reprinted from Pössinger, H. *Landwirtschaftliche Entwicklung in Angola und Moçambique* (Munich: Institute for Economic Research, 1968). With permission.

Nomadic Pastoralism

Nomadic pastoralism is another system that invests almost no labor in permanently modifying the environment, though it must be remarked that the frequent use of fire as a range-management tool over the centuries may well have modified woodland into grass savannas. This is difficult to prove, but if it is the case, then the nomads have been significant modifiers of the natural environment.

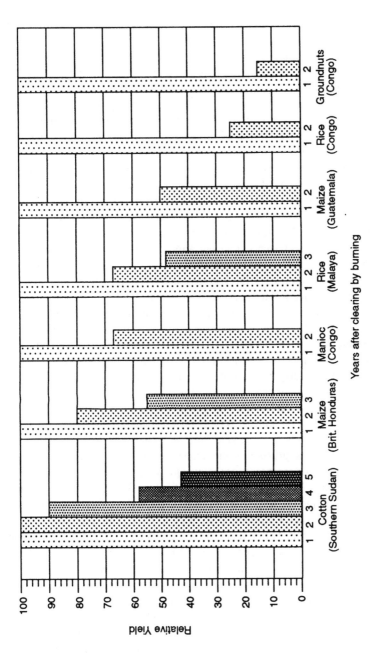

Figure 6.1. Decline in yields under prolonged cropping without fertilizer in tropical rain forest areas. (From Nye, P. H. and D. H. Greenland. The Soil Under Shifting Cultivation [Oxford: Commonwealth Agricultural Bureaux, 1965], pp. 73–74.)

The defining characteristics of nomadic pastoralism are that the animals are generally raised for subsistence, that subsistence is derived primarily from the products of the animals and not from the eating of animals themselves, that often more than one ecosystem is used, and that movement is an essential part of survival.

Nomads mostly populate the drier tropics, largely because their system of production could not compete with agriculture in terms of productivity in the humid regions, animals being secondary producers. Thus the pastoralists inhabit areas of marked seasonality of climate and that is principally why they are nomadic. The nomadic system, like shifting cultivation, has been subject to bad press. It has been described as wasteful, destructive, and subject to cultural constraints that lead to rangeland degradation and the virtual impossibility of incorporating these communities into a modern, commercial economy.

Movement, which is a defining characteristic of nomadism, allows herders, shepherds, and other nomadic groups to follow the rains. During the wet season, herders may expand over a wide domain of semiarid rangeland. The dry season requires the animal raisers to confine their activities to those locations where there is water, so concentrations of animals occur around water points. Thus the seasonal pattern is one of dispersal and concentration. Moving into rangelands at the onset of the wet season, herders may burn over grasses to bring on a nitrogen-rich flush of new grass and to control the insect population, which can carry disease. The structure of the animal population will almost inevitably reflect the subsistence requirements of the human population. Among African cattle raisers, for instance, the majority of animals will be female because the communities live from the milk supply. Young males may be slaughtered early because they compete with the females for grazing and for water, while they produce very little.

It is important, also, to consider the question of animal numbers. The most frequent complaint about nomads is that they measure their wealth in numbers of cattle and this leads to excessive herd sizes and enormous pressure on unimproved communal grazing. At the same time they are reluctant to sell their animals because cattle represent (or are) wealth. Under closer examination the "numbers mentality" becomes something different. Because of the seasonal nature of rainfall, and the perpetual risk of drought in the semiarid regions, the herder has to construct a production strategy to cope with years of extreme hardship and not just the better years. At any given moment only a proportion of the herd will be producing and part of the milk has to be fed to the calves. In addition, total yields of milk per animal are extremely low. Then again, in extreme hardship the production of milk will go down because grazing and water are in short supply and some animals may die. Even though the herder will try to minimize the number of (in human food-supply terms) nonproductive animals, it may well be the case that in hard times, even a seemingly large herd may be insufficient to provide for the herder's family or community. This has been examined quantitatively by Dahl and Hjort.[4] In this context it will be unlikely that the herder regards the family as possessing any real surplus. This would only arise where the number of productive animals exceeded the quantity considered by the herder as adequate to get through the worst of times. It is perfectly true that, within the community, someone with a large number of animals may have a very high standing, but this does not reduce the argument that the animals have a risk-minimizing function.

Within this general strategy, some communities have devised other coping mechanisms. Some may have a symbiotic relationship with neighboring farming communities, others may resort to cattle raids in time of hardship. Many of them have devices whereby animals are shared through the kinship system to reduce the risk of concentrated loss or to offset a loss already suffered. Again there is no investment of time or energy into specific land areas, rights to which are nearly always vested in the community. Despite popular ideas to the contrary, nomads almost never wander aimlessly over the land. They tend to follow well-known circuits based on an evaluation of the weather conditions in any one year. Decisions regarding these movements may be left to the elders who are the repository of environmental lore.

As with shifting cultivation, pastoralism is an extensive system. Because it exploits unimproved rangelands in dry areas, the amount of available grazing land may be limited. In addition, grazing can be exploited only where there is also access to water. Again, as with shifting cultivation, herders may spread their risks by raising more than one type of animal since different animals exploit different vegetation types (goats and cattle may both be raised, for instance, since goats browse bushes and cattle graze grass) and may represent a net increase in the supportable biomass. Goats are also resistant to diseases that might kill cattle and recover quickly through rapid breeding. Once more, the land requirements of this system are considerable, sometimes as much as 150 mi^2 per family, and so the unimproved rangelands may support only a small density of population under this form of land use. If population grows there will come a dangerous contradiction between the fact that animals are individually owned, while the rangelands are common property. This is a situation very much akin to that described by Garrett Hardin in his classic work "The Tragedy of the Commons".[5] Thus, as we shall see in the case of large families in Kenya (Chapter 13), what is individually rational as a strategy ends up being collectively suicidal.

Conclusion

These examples have demonstrated the way in which people adjusted to the vagaries of nature in particular environments: one humid, one semiarid. In both cases they maintained the natural process of regeneration, often by some process involving moving. In both cases, also, the breakdown point of the system comes at a relatively low level of population density. In the past there were many factors in the natural environment that would have acted as a constraint on numbers: disease, drought, famine, pestilence, warfare, and so forth. It would have been difficult for human or animal numbers to grow in the way that they are now able, thanks to the benefits of preventive human and veterinary medicine, pacification, water-tapping technology, etc. However attractive these systems were in the past, despite the frustration they caused the

Western mind, their sustainability is threatened by insupportable rates of population growth. As population grows, inevitably the farmers and herders must, in the absence of capital and modern technology, invest their labor in a more settled form of activity, with less return for each hour worked. This process, known as *involution* and first described by Clifford Geertz,[6] is outlined in the next chapter. The range of traditional adaptability is impressive, but it has to be recognized as a work-harder-for-less option for the person who has to undertake it. So, as population grows, it is increasingly necessary to indulge in adaptations *of* nature — which is the story of modern farming.

REFERENCES

1. Dumond, D. E. "Swidden Agriculture and the Rise of Maya Civilization," in *Environment and Cultural Behavior*, A. Vayda, Ed. (Garden City, NY: Natural History Press, 1969), pp. 332-349.
2. Conklin, H. C. "An Ethnoecological Approach to Shifting Cultivation," in *Environment and Cultural Behavior*, A. Vayda, Ed. (Garden City, NY: Natural History Press, 1969), pp. 221-233.
3. Ruthenberg, H. *Farming Systems in the Tropics* (Oxford: Oxford University Press, 1971).
4. Dahl, G. and A. Hjort. *Having Herds: Pastoral Herd Growth and Household Economy.* Stockholm Studies in Social Anthropology (Stockholm: University of Stockholm, 1976).
5. Hardin, G. J. "The Tragedy of the Commons," *Science*. 162:1243-48 (1968).
6. Geertz, C. *Agricultural Involution: The Process of Agricultural Change in Indonesia* (Berkeley: University of California, 1963).

OTHER USEFUL READING

• Grigg, D. B. *Agricultural Systems of the World* (Cambridge: Cambridge University Press, 1974).
• Hardin, G. J. *Managing the Commons* (San Francisco: W. H. Freeman, 1977).
• Nye, P. H. and D. H. Greenland. *The Soil Under Shifting Cultivation* (Oxford: Commonwealth Agricultural Bureaux, 1965).
• Pössinger, H. *Landwirtschaftliche Entwicklung in Angola und Moçambique* (Munich: Institute for Economic Research, 1968).
• Turner, L. and S. Brush. *Comparative Farming Systems* (New York: Guilford Press, 1987).

CHAPTER 7

Population and Land Use: Adaptations of the Environment

The Concept of Critical Population Density

In 1803, when most of Europe was pondering the problem of France, an English clergyman named Thomas Robert Malthus published a work entitled *An Essay on the Principles of Population.* In this book he considered the relationship between the growth of population and the availability of cultivable land to support that population. He put forward a gloomy hypothesis based on the fact that the amount of land in the world is fixed, while the human population showed a capacity for unlimited growth along an exponential curve. Prior to the time he was writing, there had been a notable famine in Sweden that had caused considerable hardship, suffering, and death.

From his observations he proposed that as population grew, most of the best land would be consumed first, and then the additional human beings would have to seek their livelihood by expanding onto uncultivated, poorer lands at the margins of cultivation. Meanwhile there would be pressure to use the existing cultivated land more intensively as numbers grew in those locations. The people moving out onto more marginal lands would find it increasingly difficult to obtain the sorts of yields supported on the better soils, and so they would need *more land per person* to support the family at the existing level. Cultivation in the poorer areas faced the risks of diminished rainfall, less resilient soils, and a weakened population that might be more susceptible to disease. As population continued to grow, the condition of the marginal population would become increasingly worse. The overall population condition

79

would then tend to correct itself by periodic famines, pestilence, and poverty, which would reduce human numbers at the margin. Malthus expressed it this way: "There is, at any time, in any given community, a warranted rate of increase [of population] with which the actual rate of growth tends to conform." This was particularly true in the time of Malthus because of the poor communication that existed then, making it difficult to move food supplies to relieve stricken areas. He appeared to predict a dismal future for mankind as it progressed toward a situation of natural correction of numbers in a harsh manner.

We may consider these ideas in the context of the system of shifting cultivation, outlined in the previous chapter, though the reasoning may be applied to any system of agriculture. If we look at the three graphs in Figure 7.1, taken from the work by Hans Ruthenberg,[1] we see in the top figure a situation in which there is an abundance of land. This means that there is plenty of time for the fallow regeneration to run its course to restoration before cultivation begins again. In the middle graph the population has grown to the point where it exactly equals the number that can be supported by that area, while allowing time for the regenerative cycle. In the lower graph the population has grown to the point where families need to start recultivating old plots *before* the cycle has had a chance to run its course. At this point, the level of fertility of the newly cleared plot is less than that needed to support the anticipated yields of the range of crops to be grown in the normal cultivation cycle. In effect what is happening in the third graph is that, due to the excessive population size, the inhabitants of the area are now *mining* the energy base of their system. In order to combat the decline of energy, the traditional farmer has two choices: (1) leave the area if other land is available or (2) farm more intensively and get a lower return for each hour worked. Since the land quality is deteriorating, as time goes by the farm family will in fact require even more land for cultivation because the yields are also declining. Unchecked by Malthus's gloomy predictions of death at the margin, the energy loss would spiral downwards at an accelerating rate.

The ideas expressed in these three graphs may also be thought of in another sense: that of *critical population density* The critical population density is the maximum density of population that can be supported on a given area of land using a given system of technology on a sustainable basis. This is, in fact, an energy relationship because the critical population density really represents that point at which, over time, the human population is balancing energy inputs with energy extraction. The energy, of course, is not total energy, but that part of the energy spectrum that is essential to maintaining the continuation of food supplies for the farmers and their dependents. It should be noted that the critical population density concept relates to three conditions: (1) a fixed area of land of a stated quality, (2) a given technology, and (3) an extended period of time. The *carrying capacity* of the land is an idea that employs the same concepts and assumptions.

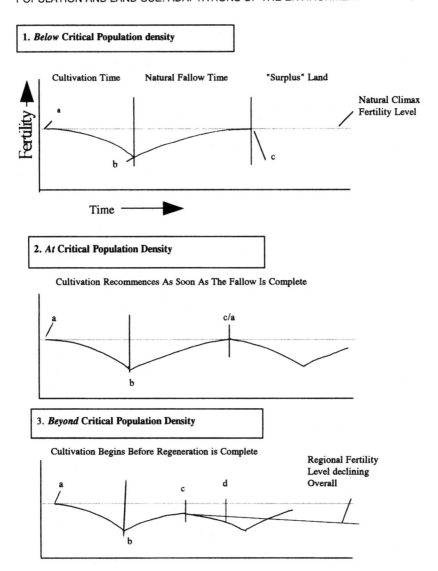

Figure 7.1.   The critical population density of a shifting-cultivation system. Key: a = cultivation commences; b = cultivation concludes, family moves on; c = natural fertility level restored; c/d = time deficit for natural regeneration to occur. (Adapted from Ruthenberg, H. *Farming Systems in the Tropics*, 1971. By permission of Oxford University Press.)

Malthus in the Tropics

Our knowledge of past demographic changes in the tropics is hazy and is rarely supported by any accurate quantitative data. It is possible to speculate, however, that the sorts of rates of population growth that we see today would have been unusual, and unsustainable, over many areas. Societies were much more isolated than they are today and so were much more at the mercy of the elements. Droughts, famines, floods, and disease would have had a more immediate impact, as would locusts and other plagues, requiring a society to fall back on its own coping devices insofar as they would carry the people. The folklore of many tropical societies is demarcated by legends of famine and pestilence (remember Joseph and his success in coping with seven lean years in Egypt). In Niger, in Africa, some groups in the drier parts of the country organize their historical calendars according to great periods of famine (the time of the great migration, the time many children died, etc.).

Occasionally it was possible for numbers to build up in excess of the carrying capacity, and then people either changed to a more intensive system of cultivation or migrated. Rarely would a people carry on in an unmodified way, steering a course toward the destruction of themselves and their habitat. Some areas, as noted earlier, were able to support large densities of population because of the quality of their soils, but this does not change the inevitable hypothesis about eventual breakdown without change. In fact these better soils often built up excessive numbers of people, in part perhaps because they were sometimes associated with volcanic highlands, which were defensible and less prone to the diseases of the lowlands from which people migrated in great waves. Other great densities of population, as we have seen, were associated with irrigation. However, in general, high infant mortality, periodic disasters, and warfare all took their toll.

An Alternative to Malthus

One problem concerning the Malthus hypothesis relates to when it was written. At the end of the 18th century, the agricultural revolution had barely begun to make its mark on productivity, though there had been some notable achievements in livestock breeding and land engineering. What Thomas Malthus could not have foreseen was the enormous change that would result from science and technology and from their application to the land. Throughout the 19th and especially the 20th centuries, the scientific community has enabled farmers to increase their land and personal productivity enormously. Today, less land is required to feed each individual. As the technology has changed (one of the qualifying assumptions of the critical population density concept), so has the carrying capacity of the land. This applies only, of course, where that technology is available. It is largely, if not almost entirely, capital based, and so access to capital is likely to be a major constraint. As is all too obvious, capital is in scarce supply in most tropical regions.

In the period after Malthus wrote, industries began to produce agrochemicals, machines, and new sources of motive power for the farm. This transformed the situation in a way that may be considered, over the long span of history, to be a quantum leap rather than an evolution. It moved agriculture onto a totally different plane of productivity by freeing cultivation from the constraints of human and animal labor and the limitations of purely local energy sources. Of course, it may be argued that this does not change the Malthus argument, it merely *delays* it, depending on one's faith in the capacity of science and technology to keep ahead of population growth and the capacity of the entire earth to absorb all the inputs without lasting damage.

A Danish writer of the postwar period, Ester Boserup, had a different view from that expressed by Malthus of the relationship between people and the land. Essentially Malthus was looking at population as a dependent variable at the whim of nature. Boserup considers population pressure, far from being the engine of a society's destruction, to be the stimulus to change. Writing in 1961, she states: "The main line of causation is population which leads to agricultural change."[2]

Boserup postulates that the stimulus of environmental breakdown and the inevitable incapacity of the existing system to go on supporting the community leads the community into new directions of land use. While this will naturally be limited by the "state of the art" of the community, Boserup demonstrates that traditional societies had far more capacity to innovate than has been realized. In Chapter 3 of her book *The Conditions of Agricultural Growth*, Boserup illustrates a succession of cultivation in which traditional societies adapted to increasing pressure. A brief summary of this succession would be

- *Forest fallow* (shifting cultivation) — 20 to 25 years for secondary regeneration; no tools other than a digging stick; use of fire
- *Bush fallow* — 6 to 10 years allowable for regeneration; introduction of the hoe as more land management required; burned vegetation for fertilizer
- *Short fallow* — annual cultivation with a few months of rest; the use of animal manure to supplement fertility; draft animal power
- *Intensive culture* — no fallow; intensive fertility maintenance and green manuring; transplanting; environmental modification on a large scale

In this way people may move through a continuum of intensification (involution). The only difficulty with this argument, which in no way undermines its essential reasonableness, is that communities end up working harder for the same reward until they enter the era of capital-intensive farming. An example of just such an adjustment is shown in the accompanying box, which describes what happened to a community, the Bakara, that was confined to an island in Lake Victoria in East Africa and where the population grew beyond the capacity of the previous bush-fallowing system.

Changes in Population Density and Pressure

The coming of the colonial powers to the tropics resulted in a number of changes in the population dynamics and put traditional land-use systems under great pressure. Although the early days of European annexation involved a great deal of warfare, the colonial powers eventually established a control mechanism that prevented much of the intergroup warfare that had existed

Text Box 7.1
The Intensive Cultivation of Ukara Island, Tanzania

The following techniques of intensification were applied to accommodate growing population density, which by the 1960s had reached over 500 persons per square mile. The Bakara produced a labor-intensive system of mixed farming:

- Grasses grown on otherwise uncultivable land and fed to animals
- Pits dug near the lakeshore for irrigation of crops
- All spare vegetation used as fodder for animals
- Zero grazing (animals kept at home and hand fed)
- Pits dug to collect all household waste, which is then carried as head loads to the crops
- A specific green manure crop grown between the crop rows and dug into the soil to help sustain cultivation
- A three-year rotation established to maintain nitrogen and prevent soil depletion
- Ridges and small stone terraces created to extend the area of cultivation

Source: Allan, W. *The African Husbandman* (Edinburgh: Oliver and Boyd, 1965).

previously. More significantly, the colonial powers began to introduce Western-style preventive and curative medicine. This was to have a dramatic impact on the survival rate of young people. Traditional communities had often placed a great premium on young people as a source of security for the parents in old age. Women continued to have large numbers of children, but now more of them survived and had children of their own. Average life expectancy increased, and famine relief assisted communities through times of hardship, adding to the total buildup of numbers. The requirements imposed on the local population by the colonial authorities to grow additional crops for taxes and to sustain the export economy further increased the per capita pressure on the land (see Chapter 8).

The importance of the colonial and postcolonial impact to our study is not just in the numbers of people that were added, but the pace at which the population dynamics began to change. It has been demonstrated that, in the past, communities showed an ability to adapt as numbers grew, but now numbers were able to grow at a speed unprecedented in traditional times (Table 7.1). Indigenous risk-reduction techniques were not evolving at an equal rate, prohibiting people from coping with the growth. Today we are used to seeing population growth figures of between 2 and 4% per annum. Taking the latter figure, the population will double in about 17 years, quickly placing the systems of cultivation under insupportable pressure. This pressure may cause the critical population density to be surpassed dangerously fast. Such dynamics

Table 7.1. World Population Growth by Decade: 1950 to 2000

Year	Population (billions)	Increase by decade (millions)	Average annual increase (millions)
1950	2.565	N.A.	N.A.
1960	3.050	485	49
1970	3.721	671	67
1980	4.477	756	76
1990	5.320	843	84
2000*	6.241	921	92

*Projected.

Source: Francis Urban and Philip Rose. "World Population by Country and Region, 1950–1986, and Projections to 2050," U.S. Dept. of Agriculture (1988).

were a feature of Europe and North America in the 19th century. There, however, wealth increased with the industrial/agrarian revolution and led to a decrease in family size (sometimes called the demographic transition). In the tropics, all too often, the growth in population exceeds the growth in the creation and distribution of wealth, and the pressure remains to have large families. This, possibly, poses the greatest threat to the tropical environment at the present time. Unless this situation changes, there is a real possibility of a Malthusian scenario involving the death of many people and the degradation of land resources. In the postcolonial period, governments have been largely unable to correct this situation and provide the levels of economic growth needed to correct the demographic explosion. This leaves us with a population/ poverty trap that is threatening the sustainability of many communities.

Alternative Pathways Following Population Growth

To put the problem in simple terms, the options facing a community experiencing a high rate of population growth are illustrated by the three graphs in Figure 7.2.

In each case, two pathways are shown for the community. One involves doing nothing, allowing population to continue to grow, and using the same system of technology. The second pathway involves working harder to maintain the same level of food output for each family.

On the first graph the impact of population growth on yield per acre is illustrated. As the community moves toward the critical population density, there is still adequate land available to maintain the system under the prevailing technology. Of course poorer land will have to be incorporated, and so yields will begin to drop before the critical density is reached (unless the land is of uniform quality). Once the critical point is passed, the yields will decline and will continue to decline at an accelerating rate if the community does not change its ways. On the other hand, if the community chooses to work harder

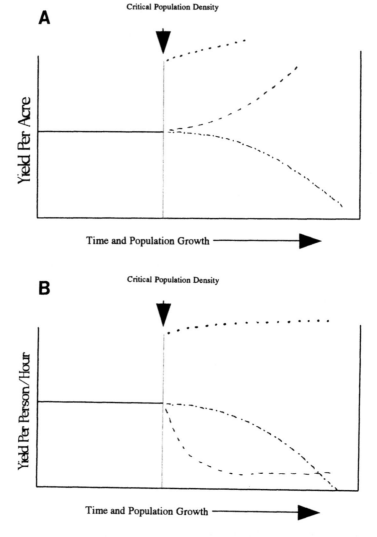

Figure 7.2. Precapital ("traditional") and postcapital responses to the buildup of population pressure. (A) Yield per acre; (B) yield per person per hour; (C) food
available per person. Key: —·— continuing old practices unchanged; —
increasing labor input; ··· the introduction of capital.

(the *involution* condition mentioned earlier) and intensify the system, as occurred
in Ukara, then the yield per acre will rise and the standard of living may be
maintained.

The second graph illustrates the impact of population growth on yield per
hour worked. Again, before the critical density is reached, the yield remains
more or less constant (apart from weather-induced differences from year to
year). Once the critical point is passed, then the yield per hour declines

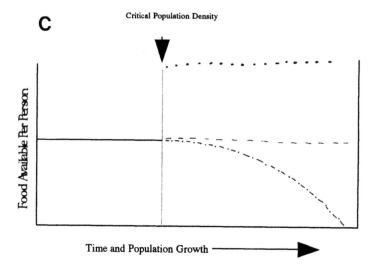

Figure 7.2 (continued).

inexorably if the population remains on the same course. Eventually the yield will fall below the point where it will feed the marginal additions of population, and the Malthusian argument will then come into play. On the other hand, if the community decides to work harder, then there will be a sudden and dramatic drop in yields per hour worked. But at least this will mean — greater effort per person notwithstanding — that the community has moved to a system that will continue to sustain itself until the critical population density for the *new* system is reached. This is undesirable, but unavoidable.

The third graph shows us the impact of an expanding populating on food available per person under the two assumptions. Again, the "do nothing" option produces a steady decrease in the standard of living and an eventual point where there is not enough food for the additional population to support itself. The "work harder" option continues to maintain the living standard in terms of food, but, as is seen from the second graph, at the expense of much greater effort. While we are often visually impressed by more intensive systems of traditional cultivation, the second graph illustrates the cost in additional human labor by which those manipulations of the environment are achieved.

All three figures show that where the Western capital option is added, there is a dramatic increase in the yield per acre, a similar increase in the yield per hour worked (because the capital is doing the work for the people), and the potential for a big increase in the food available per person through agricultural surpluses.

Options for Societies Facing Rapid Population Growth

In an article based on his work in Nigeria, Vermeer[3] illustrated the options open to societies facing the buildup of population to a critical point, as well as the barriers to achieving these options. His diagram includes some options that are of a decidedly"nontraditional" form relating to a more modern, though not yet "developed", economy. A much modified version of his diagram is shown here as Figure 7.3. The first of these options is the most obvious, which is to bring more land into production. Clearly the best and most easily used land will have been cultivated first, and thereafter it becomes progressively more difficult to clear or use land, and sustain production from it. Generally this option is going to require increasing amounts of energy per marginal unit brought into production. There may simply be no more usable land left in the community's domain once terracing, irrigation, drainage, and all other landscape-modifying options have been tried. Or, there may be severe sociolegal barriers closing off access to land even by members of the same nation or society, through unequal tenure systems for instance.

Option two involves increasing agricultural productivity to feed the population (more output from the same amount of land). This will be constrained by the technology of the society (what options they know of or can devise), the availability of capital, and physical factors such as those limiting irrigation, terracing, etc.

A third option — in more contemporary times — would be the adoption of nonfarming activities. Traditionally this option was limited by the market and the ability of agricultural production systems to feed nonfarming specialists. Today this pathway frequently involves the drift to the towns: an action that is severely constrained by the chronic unemployment situation in many tropical, developing countries and again by the limited size of the domestic market to support manufacturing and service activities.

A fourth option is the formerly widespread one of horizontal expansion into the wilderness, or taking in additional land through migration, warfare, and raiding. The creation of Western-style settled administrations and boundaries in tropical areas severely limits the possibility of wholesale movement or migration and imposes central control over warfare. External migration is limited by the fact that countries do not easily accept unskilled migrants from poor countries, since they have problems of a similar type at home to deal with. However, some countries, such as Somalia, have more of their people abroad than they do at home and the remittances of these emigrants play an important part in sustaining what is left of their home country's economy. Much the same may be said for Pakistan and Bangladesh.

An additional option is to limit fertility within the community so that fewer children are produced, thereby reducing the buildup of pressure on the land. This might seem a logical step because it attacks the root of the problem, but as we shall demonstrate in the case of Kenya, there are problems of survival and the concept of individual-versus-collective rationality. It may not make sense to the individual family to reduce the number of children it produces. Some community groups, such as the Kalahari Bushmen, are known to have practiced fertility control. The problem is not so much a cultural as an

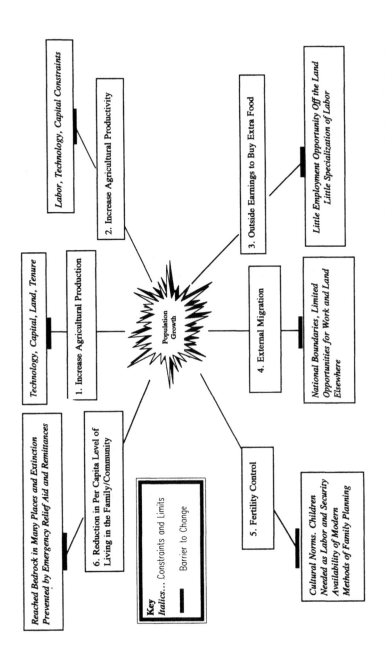

Figure 7.3. Opportunities and limitations relating to population expansion in a traditional agricultural community. (Adapted from Reference 3.)

economic one, and much of the difficulty again arises from the fact that the speed at which population now grows has overtaken the rationale of earlier systems without producing a viable alternative.

A final alternative is for growing populations to accept a declining standard of living. Many communities have had to do this, especially in Africa, although it is difficult for those communities that are already close to subsistence. In recent years this option has been held at bay, or at least reduced from achieving totally catastrophic results by the massive inflow of remittances and foreign relief aid. Unless this generates systemic change it simply allows the community to buy time while the problem gets **much** worse.

Conclusion

It is now very clear that many fundamental changes have occurred in the relationship between people and land since agriculture began. These have allowed us to feed unimaginable numbers of people. The ways in which our symbiosis with the land has changed are shown, in a much simplified form, in Figure 7.4. Even though there are vast geographical variations in how far different societies have moved along this curve, there has been an unquestioned orthodoxy that this was the "way to go". Such slavish copying of the past maybe answers the question "How will we ever feed so many people?" It does not answer the question "How much of this will the earth tolerate?"

Although many traditional societies have shown a considerable capacity to adapt to population change in the past, the problem is now one of an accelerating rate of change (see Table 7.1), threatening us with real and potential densities of population that are simply unsustainable. As we shall explore in the case studies, the population/poverty trap is unlikely to be resolved simply by attacking human fertility rates and nothing else. A way has to be found to produce sustainable land-use systems and sufficient security to reduce the *perceived necessity* to have so many children. The traditional societies in the tropics have had to deal with a situation in which their circumstances did not evolve in the way they did in the West, but were subject to a sudden and unequal disjunction that is the subject of the next chapter. This historical reality must be understood and lies at the base of any proposed solution.

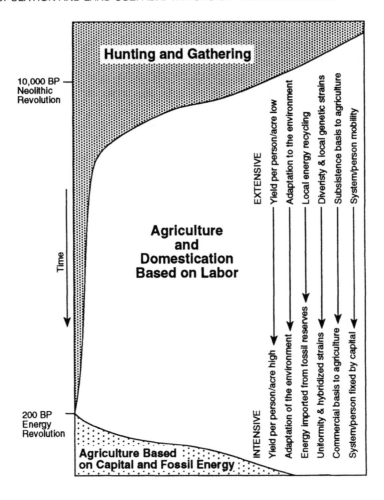

Figure 7.4. People and the land — 10,000 years of change.

REFERENCES

1. Ruthenberg, H. *Farming Systems in the Tropics* (Oxford: Oxford University Press, 1971) p. 47.

2. Boserup, E. *The Conditions of Agricultural Growth: The Economics of Agricultural Change Under Population Pressure* (London: George Allen and Unwin, 1965).

3. Vermeer, D. E. "Population Pressure and Crop Rotational Changes among the Tiv of Nigeria," *Ann. Assoc. Am. Geogr.* 60(2):299-314 (1970).

OTHER USEFUL READING

- Allan, W. *The African Husbandman* (Edinburgh: Oliver and Boyd, 1965).
- Brown, L. R. and J. L. Jacobson. "Our Demographically Divided World," Worldwatch Paper #74, Worldwatch Institute, Washington, D. C. (1986).
- Ehrlich, A. H. and P. R. Ehrlich. "Why Do People Starve?" *Amicus* J. 9(2):42 (1987).
- Hedrick, P. *Population Biology: The Evolution and Ecology of Populations* (Boston: Jones and Bartlett, 1984).
- Mamdani, M. *The Myth of Population Control: Family, Caste and Class in an Indian Village* (New York: Monthly Review Press, 1972).
- McKeown, T. R., G. Brown, and R. G. Record. "An Interpretation of the Modern Rise of Population in Europe," *Popul. Stud.* 26:3 (1972).
- Preston, S. H. "Population Growth and Economic Development," *Environment* 28(2):6 (1986).
- Repetto, R. and T. Holmes. "The Role of Population in Resource Depletion in Developing Countries," *Popul. Dev. Rev.* 9:609 (1983).
- Sai, F. T. "The Population Factor in Africa's Development Dilemma," *Science* 226:801 (1984).
- Urban, F. and D. Rose. "World Population by Country and Region, 1950–86, and Projections to 2050," U.S. Department of Agriculture (1988).

CHAPTER 8

Europe Annexes the Tropics

The tropical societies and the agricultural systems that supported them underwent a dramatic period of shock and change from the end of the 15th century onward. This shock was connected with the expansion of Europe and the growth of the colonial economy, which was to shake the so-called "traditional societies" to their foundations. To understand what happened during this period of contact and conflict, we need to understand not only the systems that the Europeans found (as discussed in Chapter 6), but just as importantly, what the Europeans thought of these systems of production and what the Europeans were trying to achieve by establishing colonies in the tropics and subtropics. We tend to look at what happened at the time of European intervention from the point of view of a 20th-century person with 20th-century knowledge, values, and information. Instead, we need somehow to get inside the minds of the Europeans of the 17th, 18th, 19th, and even early 20th centuries. Only by making this adjustment can we begin to understand what the Europeans were setting out to do and how they viewed the world at that time. We have already examined the basis for the rationale explaining the indigenous systems. It is also important to remember that these ideas changed as time went by and, indeed, are still changing.

Why Europe Expanded into the Tropics

While it is quite common now to think of European expansion as beginning with the journeys of the Portuguese around Africa and Christopher Columbus's voyage to the Americas at the end of the 15th century, Europe had actually been a continent of movement, expansion, and contraction for centuries. From Italy the Romans took their empire to the far corners of the Mediterranean. Even then they were often following in the footsteps of earlier Greek colonizers who had traveled and established colonies around the same area and sent their armies as far as Persia and India. In most of these cases, however, the colonizers were seeking out lands that were, to a large extent, similar to those they left behind in their home regions of Italy or Greece. It is true that they sometimes ventured into drier lands, as in North Africa, and built on existing irrigation systems or developed new systems based on their "home-grown" models.

As far as this book is concerned, the main phase of expansion was that which took place by sea and took the Europeans into very different environments and social systems. Initially this began with a search for the riches of the Indies. In the Europe of the 15th and 16th centuries, many items of considerable value came from China and the East: spices, silks, and other precious commodities. This trade was controlled by the Turks, whose empire lay across the great trade routes from the East and across the narrow neck of land that divided the Mediterranean from the Red Sea. At that time the Europeans had no idea whether the continent of Africa was something that could be sailed around or whether it was joined to a great "southern continent", and so the overland route to the East was the only one known to be feasible. The growing merchant class resented paying vast profits to the merchants of Venice, who were the only people with whom the Turks would trade, and so, clearly, there was a fortune to be made for anyone who could find a way around the Turkish Empire and capture the spice trade at its source (approximately where Indonesia is today). The spices were of particular interest because the agriculture of the Middle Ages in Europe had not discovered how to keep large numbers of livestock alive through the hardships of winter when there were no crops to graze. Consequently meat had to be preserved in salt and quite often went bad. In these circumstances, spices could hide a multitude of bad tastes and disguise a meal so that it was fit for the tables of the rich.

The Portuguese attempted to get to the Spice Islands by going around Africa; the Spaniards took the other option and headed west to try to sail around the globe. By the first years of the 16th century, a passage round Africa had been opened and the first landings in South America and Central America had been made. In point of fact the Portuguese had little impact on the territories they annexed since they tended to restrict themselves to ports on their way to India and the Indies. The Portuguese presence in Brazil follows, reasonably closely, the pattern set by the Spaniards in the rest of Latin America, and so it is the Spanish story that we shall follow.

Spain in Latin America

The discovery of the Americas had been something of an inconvenience to Spain, since the land stood in the way of ships sailing to the Indies. But this disappointment soon changed when the Spaniards discovered the gold used by the rulers of the empires existing at the time. The widespread destruction — of societies, farm systems, and huge numbers of Indians — that resulted from the "gold fever" needs some explanation. We have seen earlier that these Indian societies had reached a high level of engineering, art, and land use, and yet these societies were utterly destroyed by the Spanish presence. Why?

Spain in the 15th and 16th centuries was gripped by a religious fever that eventually took the form of the Inquisition. The Catholic religion was the only one acceptable in the eyes of the Spanish church and court. Unbelievers, even those following other forms of Christian belief, were tortured into changing their faith back to Catholicism. If they did not change, they died, and this was considered to be "better for them" since it prevented them from continuing with a life of sin. As far as Spain was concerned, the Incas, Mayas, Aztecs, and others were all "heathens" and the mission of Spain was to convert them. Little to no value was placed on heathen life, art, or skill since, to the European mind, it was the worthless creation of unbelievers. If we accept that the Spaniards arrived with this view of the world then it is hardly surprising that they were not interested in the achievements of the Indians, only in their conversion to Christianity. In 1503 Queen Isabella of Spain wrote to her authorities in the Americas with the following instructions regarding the Indians. The governors were to "compel the Indians to be associated with the Christians . . . and to work on their buildings, and to gather and mine the gold and other metals and till the fields and produce food for the Christian inhabitants. . . ."

To put it bluntly, to the European mind it was inconceivable that there was anything in a heathen society such as this that would be worthy of study. The Indians had to be taught European values and ideas before they could advance.

We should also note that the Indians were to grow food for the Christians and this is the first recorded impact of the colonial system on the agriculture of indigenous peoples in the tropics. The food was to be grown in addition to the food that the Indians had to grow for their own needs; there was to be no choice in the matter, and it had to be given to the Europeans as a "tribute". This marks the beginning of colonial taxation. It is true that the Indians had to grow produce for their Inca or other masters before the Spaniards came, but what was now being imposed by the Spaniards was a tax to a totally outside power with no obvious return. The labor that had to be given for mining was another form of tax, and again the main difference was that the wealth was, henceforth, going to be taken away from Latin America to benefit Spain: Resources were to be drained away. In one extreme case the Spaniards totally depopulated the Bahamas in search of slave labor for the mines of Peru.

Thus the Spanish aim was simply to extract the wealth from South America, offering in return the "benefits" of Christian civilization and redemption for the souls of millions of heathen Indians. To perform this task the Spaniards used soldier-adventurers (*conquistadors*) who, in return for gaining total control over the territories for Spain, would receive huge tracts of land (some of several hundred square kilometers). To gain control these adventurers had to completely smash the power of the traditional rulers, and this they did using the superiority of their weapons. With horrifying swiftness the empires fell, and along with them much of the indigenous knowledge was lost or devalued, as was the social control that had held together the organization necessary for the terraces, irrigation channels, roads, and public safety. In the bloody wars and the enslavement that followed, the population of South America was reduced from an estimated 50 million to around 5 million. It became increasingly impossible to maintain all the intricate stonework of the terraces, and they started to collapse or were just abandoned. Today vast areas of Peru and Ecuador reveal the outlines of terraces that once raised crops, but now in ruins, having been productive in some instances for over 2000 years.

Originally, the vast grants of land made to the soldiers of fortune who came from Spain remained the property of the Crown, but the soldiers were allowed to hold onto them as long as they undertook to convert the local population to Christianity. Fairly quickly this turned into private ownership, and the pattern of land tenure so typical of much of Latin America was established. The colonists naturally took the best land for themselves while the peasants who lived on it were required to perform certain tasks — another form of taxation — in a sort of feudal obligation to their new masters (*peonage*). As the best land became swallowed up into *haciendas*, often very lightly used or used for the raising of livestock (including cattle and horses first brought to the New World by Columbus on his second voyage), the peasants had to move out toward the drier lands or, usually, up the hillslopes. Now there was nobody to coordinate the labor force to build terraces or to design irrigation systems, and so the peasant farms were often carved out of the hillsides by cutting down the forest cover. This was to prove disastrous since it led to the sudden exposure of these previously well-protected hillslopes to soil erosion. Thus the typical upside-down rural population pattern of Latin America developed with the high densities of rural population squeezed onto the poorer lands and the better land being under low-density use or, in some cases, not used at all, but simply kept as an investment or a hedge against inflation.

The Expansion of Britain

Britain's attention had been focused on two main areas: one temperate (the American colonies) and one tropical (the West Indies). With the expansion into the tropics, the Europeans were to discover new plants and animals that were to have a considerable impact on trade and tastes. One of these was sugar, first grown commercially by Europeans in Hispaniola in 1509, around which almost the entire history of the West Indies was to be written. Commodities such as sugar, tea, indigo, and tobacco were primarily luxury goods and commanded extremely high prices, especially among the growing middle classes: the burghers of the expanding European townships. The high price the public was willing to pay encouraged the rapid expansion of these crops, so that by 1744 the value of England's trade with the West Indies was greater than that with the North American colonies.

Sugar cane was ideally suited to the West Indies because of its requirement for hot, wet conditions. However, its cultivation, especially its harvesting,

Text Box 8.1
Inside the Mind of 18th-Century Britain

One of the most remarkable books ever written, now almost totally neglected, was Sir James Steuart's *Political Œconomy,* published in 1757.[1] This book, written before that of Adam Smith, outlined the theory of comparative advantage and pointed out the technological solution to Malthus's population dilemma years before Malthus published his book. Steuart gives us a sharp insight into the perspective of the European mind regarding trade and "unsophisticated" nations. Here are some extracts from this truly remarkable book.

On drawing the backward nations into world trade: "The traders will, therefore, be very fond of falling upon every method and contrivance to inspire this people [the "savages" overseas] with a taste of refinement and delicacy. Abundance of fine presents, consisting of every instrument of luxury and superfluity, the best adapted to the genius of the people, will be given to the prince and leading men among them. Workmen will even be employed at home [in England] to study the taste of the strangers, and to captivate their desires by every possible means. The more eager they are of presents, the more lavish the traders will be in bestowing and diversifying them. It [the unsophisticated country] is an animal put up to fatten, the more he eats, the sooner he is fit for slaughter. When their taste for superfluity is fully formed, when the relish for their former simplicity is sophisticated, poisoned, and obliterated, then they are surely in the fetters of the traders, and the deeper they go, the less possibility there is of their getting out" (p. 185).

On why trade with individual areas should be kept exclusive, which would ultimately provide the rationale for colonialism: "When companies [monopolies] are not established, and when trade is open, our merchants, by their eagerness to profit of the new trade, betray the secrets of it, they enter into competition for the purchase of the foreign produce, and this raises prices and favours the commerce of the most ignorant savages (p. 187).

requires a great deal of labor under tiring physical conditions. Not surprisingly, the local population was not particularly interested in growing what was for them a largely useless crop. The Europeans had the lure of profit, just as the Spaniards had had through the acquisition of gold. But the Indian population had little or no desire for cash since it had never operated within a cash economy and since there was little they could do with the cash anyway. So we find an early conflict between the *commercial* nature of the colonial system (i.e., the profit motive) and the *social* nature of the precolonial systems. By this we mean that if the Indians had to produce a surplus of crops for their Inca masters, they did so because it was their social obligation to do so. They may not have liked it, but that was the way that society was organized in the Indian system: society and economy were inseparable. Most of the craftspeople producing things for sale had previously been catering to the powerful, who could afford to keep the skilled craftspeople by paying them with gifts of food, cloth, etc. which had been extracted from the poorer population as "tribute".

In these early cases, the Europeans did not waste time trying to introduce the local population to the money economy, so as to encourage them to grow the crops that the Europeans wanted for trade. Instead they tried to force the local population to work as a form of "tribute", but the rigors of the sugar fields were too much and the forced labor of the Indians resulted in their deaths in large numbers, just as was happening in the mines of Peru. The European response in the West Indies was to introduce two innovations that formed the backbone of the colonial system for many years: the plantation system and slavery.

The Plantation

The plantation system was the most obvious intrusion of the commercial colonial system into the subsistence-based indigenous system. Just as Europe was beginning to discover the idea of the factory — i.e. bringing together a lot of skills in one place, usually around an energy source or a piece of specialized machinery (capital), to produce one thing in great quantity — so the idea spread to agriculture. Previously, tropical agricultural systems had been characterized by their great variety of crops on each farm, except in the rice-producing systems of Asia. Now the Europeans were often interested in growing only one crop and growing it continuously year after year for profit. This was the idea of **monoculture**. Where it had been previously tried by Indians, it had some-times proved disastrous, as, possibly, with the Mayan maize system in Central America. But the plantation was to be a great enterprise, growing one product and **processing** it. Thus, at the center of most plantations was a factory for refining the sugar, boiling the latex, and so on, then producing the various by-products such as refined sugar, jaggary, molasses, rum, copra, etc. This factory required **capital** and **engineering skills**, which meant that it was unlikely to be developed by the indigenous population in the short run. The factory required a steady flow of raw materials to ensure that the machinery was run efficiently. If it lay idle at any time, then it was not earning money or paying for itself. It seemed almost impossible to the European mind that the local farmers could (1) understand how to grow sugar, tea, or other cash crops efficiently; (2) produce the crop at the right time for the factory to be "fed" regularly; (3) develop a sufficient interest in making money when they were such newcomers to the money economy; or (4) be prepared to produce enough of the commodity for the money being offered.

The plantation involved acquiring a large tract of land by one means or another, planting it with one crop, managing the crop centrally through skilled overseers, having total control over the labor force, and maintaining the quality of product required by the main market — Europe. So land was alienated into foreign hands; the crop was selected. That left the problem of labor. Slavery was the obvious answer since it bypassed all the problems of local farmers who were not interested in money as well as the problem of total control over the supply of the crop and it eliminated the need for cutting into profits to pay other farmers to grow the crop and supply the factory. In this way, the indigenous people became a "factor of production", subject to total control. In 11 years (1640 to 1651), the slave population of Barbados grew from 200 to 20,000.

The initial success of the plantation system and the huge profits made from sugar by the colonial settlers encouraged the spread of the system to a wide range of crops. In south India and Ceylon (Sri Lanka), the plantation system became the basis of a vast British-owned tea operation employing thousands of laborers on subsistence wages. In Southeast Asia, the plantation system was extended to the cultivation of rubber on European-owned estates using coolie labor.

As use of slave labor came to an end, Europeans sought to find some method of involving the local cultivator in the production of the crops demanded by the hungry industries of Europe: cotton for the spinning mills of Lancashire, vegetable oils for lubrication and candles, and coffee, which was rapidly gaining popularity. Even after slavery was abolished, it was possible to continue with "disguised" slavery such as the labor indenture system that was used to ensnare Chinese and Indian (i.e., from the Indian subcontinent) laborers, as on the French island of Réunion or in British South Africa or Fiji. Increasingly, however, it was clear that some means had to be found of "transforming" traditional agriculture in order to bring the peasants into the commercial economy. This was particularly a problem for the colonial powers in Africa and parts of Asia. In Latin America, the old system of using the peasants as bonded labor on the farm was much preferred by those in power, and little official attention was directed toward the peasant farmers as long as they turned up in sufficient numbers to work on the *haciendas* or on the largely foreign-owned plantations.

Commercializing Peasant Agriculture

So far what had been created in the dependent countries was what has been called a "dual" economy. By this it is meant that there were islands of European investment (the plantations mostly) surrounded by largely unchanged subsistence cropping. While this may be true in a general sense, we must not conclude that nothing ever happened to the local agricultural systems. We know, for instance, that there was a massive transfer of crops from South America to Africa and Asia as a result of the slave trade and other connections. Maize, after all, is a South American crop, but it is widely established as a staple subsistence crop in some parts of Africa, such as Kenya. Tobacco was grown in India by 1616. There was little for the local farmers to learn from the plantation economy since it operated on a scale that they could not possibly copy. Indonesia, or more particularly Java, is often cited as the classic case of a dual economy. The Dutch carved out plantations for sugar and other crops for export while confining the Javanese peasant population to an ever-decreasing area of land. The result was an intensification of the traditional farming system to an incredible degree, providing us with the classic example of a process referred to in Chapter 7 as *involution* (i.e., turning inward on itself instead of evolving into something different).

But the colonial economy was interested in more than just extracting the wealth of these territories. It also looked upon them as great potential markets for the products of Europe. Of course nobody could buy these products without having money. So it became increasingly attractive to encourage the "commercialization" of activities in the colonial dependent territories. Where there already was a sense of money and commercial system, as in India, then the East India Company set about destroying local competition, since the colonial system was, from the beginning, essentially an *unequal* system: hence the cutting off of the thumbs of Indian weavers. Commercializing the local agricultural community and system could be achieved in one of several ways. In India one simply supported and extracted wealth from those who were already commercially minded, had power, and who could, through the traditional system, apply pressure on others more powerless below. Thus the *zamindars* (originally a form of tax collector) or landlords flourished in British India as greater encouragement was given to the commercialization of cropping and the ownership of land under freehold title. In many communities this concept had not existed previously, but had accompanied the importation of Western values along with the colonial system. Once land can be owned, it can be used as security; debtors then lose their land and become part of the army of the landless ready to work for the landlord on his commercial farm for subsistence or a subsistence wage.

In Africa the problem was different since there was no basic money system to build on in many cases, though internal trade was often well developed. The method of transforming traditional agriculture in this case was the institution of taxes, which had to be paid in cash or in *selected* agricultural products. The peasants were thrust into the cash economy and into growing crops in which they had previously had no interest and which they could not eat. It was in this way that the British introduced cotton into Uganda, suddenly issuing an order requiring everyone to pay a poll tax or hut tax. This created an immediate need to obtain cash or otherwise spend some time in jail. By 1936 Uganda had 1.5 million acres under cotton simply as a result of the introduction of compulsory taxation. But it should not be imagined that the colonial authorities wished to *replace* the subsistence economy with the cash economy. Their ideal was to graft the cash economy onto the subsistence economy for several reasons. First, it was feared that if the local population shifted entirely to cash cropping, there would be enormous problems feeding the people and that the potential for famine would increase due to the poor food-distribution system. The British had sufficient experience with famines in India to realize the importance of maintaining an adequate food-producing base in the new African territories. Second, it was realized that when Africans were being drawn out of the countryside to work for the Belgian, French, British, or other colonial masters, then, if the remainder of their family stayed behind on the farm and fed itself, the wage paid to the man employed could be reduced by that much. This was critically important in places such as Northern Rhodesia (Zambia), where thousands of laborers were being extracted for the mines, and in the Congo Free State (Zaïre), where they were being put onto plantations.

So a new type of "dual" economy was encouraged, this time with the twofold division occurring on the farm itself. The cash crops were grown in addition to the more usual food crops; the two economies, eras, and worlds existed side by side. This led to an immediate increase in the demand for land since the farmers had to increase the scale of their activities in order survive in both worlds at once. We have discussed the impact of the colonial period on population dynamics in Chapter 7.

At the time the colonial authorities were unleashing these great forces that have come to dominate so much of the literature about development, they were effecting other changes that were to make the rural situation even worse. Traditionally in times of population stress, people in Africa and Asia resorted to mass migration, thereby easing the pressure. With the coming of European power, such movement was discouraged. Sometimes the possible line of movement was broken by new boundaries drawn in some office in Berlin or London, boundaries that became territorial frontiers. Sometimes these lines divided communities in two, as was the case with many parts of the border between Uganda and the Sudan. People were encouraged to settle so that they could be better controlled. In particular the nomadic people who relied upon movement for their survival were constrained by the drawing of boundaries. Sometimes these made their seasonal migrations impossible. At other times their dry-season grazing was encroached upon by cultivators in need of land, as happened in the French colonies of Niger and Mali or in Rajasthan in India. In the disputes that arose between the cultivators and the pastoralists over rights of access to pools or rights of passage through crop lands, the colonial authorities virtually always sided with the cultivators and regarded the nomads, in the words of an agricultural commission in India in 1890, as "brigands and madmen". Thus old relationships were overturned and destroyed.

Adding to the boundary problem, a continuing pattern of land alienation compounded the pressures on the local farmers. This was exhibited in Spanish America in a pattern that persists today. It also became the pattern during the eras of the plantation and of the "settler". During the 19th and early 20th centuries, the European expansion was in full steam. Millions of poor Europeans streamed into the United States, Australia, Canada, and the tropical dependencies. The difference between the first three and the last is that the population density of indigenous peoples in the first group was slight, so that these people were displaced or destroyed with relative ease. In most cases in the tropical dependencies, however, there was the complication — from the colonial point of view — of millions of indigenous farmers or herders to deal with. Total white settlement was not possible, and most European settlers preferred to go somewhere that was environmentally more familiar anyway. In Kenya or Algeria, for instance, large tracts of land were excluded from native settlement and handed over on a freehold, or long leasehold, to European settlers. The local population was then confined to defined areas, usually called reserves, much as the American Indians were confined to their reservations by the new immigrant government. Usually it was the best land that was expropriated and handed over to the Europeans.

The creation of these reserves added one more pressure factor to those already noted — accelerating population growth and the rise in the demand for land per family for cash crops. Soon the first signs of serious erosion became noticeable in those areas designated as reserves. This was the same problem, of course, as the situation of soil degradation and erosion on the hillsides of Latin America or on the over pressured terraces of Java where the population had been confined. To some extent the soil erosion was accentuated by the technologies that the Europeans brought with them.

It has to be remembered that those Europeans who moved into the tropics took with them the ideas and techniques that had been developed in Europe under European physical conditions. Indeed, most of them were developed in northern Europe, where the soils are deep, well structured, and able to absorb considerable amounts of abuse and intensive use partly because the land is rarely left bare and because the rainfall is gentler and fairly well distributed. The techniques suited to Europe were carried to the tropics with little or no modification and were often applied to new crops such as maize or cotton. The Spanish plough, for instance, was introduced into South America with disastrous consequences since it turned the soil and dried it out, leaving a dry tilth that was rapidly blown away. This same moldboard plough, incidentally, was to have equally disastrous consequences in the American Great Plains and in South Africa. Earlier indigenous methods — using digging sticks or simple ploughs that scraped rather than turned the soil — disturbed the soil to a much lesser degree.

Where the Europeans, or those of European descent, observed the breakdown of traditional farming methods, their explanation was usually to blame tradition rather than to look to the consequences of the colonial system for any explanation. In general the stress on the land was explained by "overpopulation". In their view there were, quite simply, too many native people and they were overtaxing the land. This basic situation was accentuated by the fact that the local people continued to use traditional farming methods such as shifting cultivation which were considered to be wasteful and which, under population pressure, quickly broke down. Yet we have seen in Chapter 5 that quite often these "traditional" systems were capable of a remarkable degree of adaptation to population pressure. So why were they collapsing so rapidly in the early years of the 20th century when the agricultural and colonial journals were full of fearsome accounts of the widespread threat of soil erosion?

During the colonial period, changes occurred or were introduced so rapidly that society simply could not adjust to what was happening. We have already observed that land was alienated, medical improvements were introduced, and demands were made for increased amounts of land for cash crops — all of this occurring *simultaneously*. For many people the only answer was to move out of agriculture totally and join the army of the unemployed in the sprawling urban slums or to try to find a piece of remaining land somewhere. Thus the advances into the forests continued, and agriculture moved up the hill slopes and out toward the deserts. The colonial authorities reacted by imposing legislation on land use restricting the types of crops that could be grown or imposing upon the local population responsibilities under law to see that the land was not damaged. In Rwanda, the Belgians insisted that each community devote a number of weeks each year toward the construction and maintenance of terraces to save the hillsides from being washed away. For political or economic reasons the colonial authorities rarely agreed to release any of the alienated land, which would have taken the pressure off the indigenous peasant farms. But what value did that economy have for the small farmer who did not have enough land to grow cash crops anyway?

In short, the commercialization of the peasant economies of the tropics was part of a broad process of incorporation of these areas into a new social, political, commercial, and cultural world order. In terms of our primary concern, the management of the physical environment, the "West" was the harbinger of a new era of dysfunction with nature. Table 8.1 summarizes the ways in which the new technical, commercial order was moving away from the principles of natural, ecological order. The trend was to move from what's in the left-hand column to what's in the column on the right.

The Colonial Context

From the examples studied it is now possible to see the broad context in which colonialism operated and to draw some conclusions that explain the way it came to change the world of the tropics forever 500 years ago. In the West there is an integrity and an evolutionary continuity to history, despite incursions by Turks, Mongols, and Arabs, and despite those sudden jumps in development such as the Black Death or the Industrial Revolution. Everything interacted in one place, so society changed with economy, with technology, with culture, and so on. However, the sudden and explosive expansion of Europe into the tropics brought into the indigenous societies and their production systems a swift and terrible **quantum jump**, or dysfunction, in their history. In effect their history stopped evolving at that point and shifted to a totally different and largely nonindigenous track. Hence, what is seen today is an often unhappy hybrid of home grown and imported, perhaps best exemplified in Latin America, where Indian civilization does not sit well with the Ladino culture.

Why exactly Europe, rather than some other part of the world, broke onto the global scene and took over almost all the tropics for the best part of 300 years is difficult to explain. Perhaps it had something to do with the essentially nonpantheistic nature of the Christian religion, which perceived humankind as

Table 8.1. The West Moves Away From Nature in the Age of Science

The working of nature	The working of humankind
Works in a cyclical way, reprocessing and recycling to produce nothing that can be called "waste".	Works linearly, transforming materials into nonlasting objects and producing much waste, including heat.
No great excesses. Natural forces tend to work against excessive concentration of energy.	Large excesses in terms of huge numbers of few things (e.g., monocultures).
Competition and cooperation abound to produce dynamic stability of ecosystems.	The "conquest" of nature by Western science and technology is an effort to override the natural system.
Increases biological diversity essential for competition and survival of the fittest.	Diminishes biodiversity, encouraging "crops" and eliminating "weeds" and "pests".
Works toward global stability over time, though not necessarily toward the ultimate betterment of humankind.	Human activities are increasingly unstable in terms of locally operating ecological forces, thus encouraging further intervention.
Multiple feedback controls are mostly negative (i.e., self-regulating toward stability).	Little feedback control and mostly positive, meaning we must ultimately control ourselves or face extinction, which is the ultimate negative feedback.

Source: Loening, E. L. "The Ecological Challenge to Growth," *Development* 3(4):48–54 (1990).

dominant over nature. Once this is tied to freedom of will, and the spirit of the Reformation, the way is laid for the development of science. Previously, the "hand of God" was sufficient explanation for any phenomenon, as Gallileo found out. While many of the world's religions retained fatalism, they still seemed able to make remarkable advances, such as occurred under the Arab kingdoms. Perhaps the explanation is that the Europeans harnessed the weapons technology of the age of science, and set out to build their empires by sea, capturing wealth from a vast area.

Whatever its cause, the European imperial expansion resulted in a new world order — one that was both *unequal* and *unidirectional*. From the beginning the philosophy was one of *dominance* and *superiority*, and even Columbus set out not just to trade, but to change the religion of the people he found and to claim land for the throne of Spain. The next 300 years were to see the imposition of European values and ideas across the board. By unidirectional it is implied that those who went to the tropics as part of the imperial mission went there to change the place, not to learn from it. Such learning as there was was confined to researching the suitability of lands for colonial cash crops, a "curiosity" about quaint customs, or a superficial survey of social control to see what could be incorporated into the imperial system of indirect rule, thus cutting down on administrative overheads. The attitude of the whites to the aborigines of Australia, who were hunted down like animals, or toward the "Indians" of North America, who were systematically robbed and cheated to the point of ruin, is fairly representative of a unidirectional mind set. There was simply no interest in finding out about them, or affording their cultures and practices any worth whatsoever. What compounded this clash of cultures was the terrible *suddenness* with which it happened, not allowing any possibility for adaptation.

It is interesting to ask why the Europeans found the lands, or more precisely the people, they conquered to have no values worthy of serious study. The answer lies in the four main elements that made up the cultural context of colonialism. That cultural context is best illustrated by the language that contemporaries used about the situation, and we have seen a little of this in the earlier boxed quote by Steuart. The Victorian lexicon of colonialism is full of words such as "primitive, backward, unenlightened, savage, barbarian, uncivilized, primeval", and so on. These words all convey an attitude of evolutionary superiority, i.e., the other people are further back along the universal road to progress. Hence the use of words like primitive and uncivilized imply that there is but one type of civilization. Perhaps the best example of the choice of words can be seen in this extract from a best-selling book of 1889 by an engineer, Henry Drummond, who, it must be remembered, *loved* Africa:

> Hidden away in these forests like birds' nests in a wood, in terror of one another . . . are small native villages; and here in *virgin* [here is another "unevolved" word] simplicity without clothes, without civilization, without learning, without religion — the genuine child of Nature, thoughtless, careless and contented.[2]

The simple fact is that these people had *different* forms of civilization and learning, but Drummond was not ready to admit that they could have any validity. Nevertheless, reading a piece like this helps us understand what happened. Furthermore, it should be pointed out that Africa had been ravaged in the search for slaves for several hundred years by the time Drummond arrived on the scene. It is hard to hold any sort of civilization together in those circumstances.

The principal components of the unequal, unidirectional approach were religion, science, and economics.

1. *Religion* — The entire thrust of the Christian mission was to save souls and convert people, and this does not allow for other religions to have much validity, or as Drummond points out, even to be considered religions at all. At its extreme level, under the Catholic inquisition, the Christian missions offered death as the alternative to conversion. Another popular word in the colonial lexicon was "heathen", which implied the Godless nature of non-Christians. In fact Christianity was to replace the entire traditional constellation of gods. But those traditional religions often bound people together, established their relationship with nature, and so forth. So replacing the old faiths was about the most disruptive thing that could be done to the old ways. The religion was the glue that rationalized everything, made something good or bad, and allowed one to judge one's peers. This was not only

smashed, it was held up as something *worthless*, making people ashamed of their culture, their history, and their worth. This transformation was quite possibly the most damaging phase in the history of the tropics since it cemented the relationship of inequality to the new masters. The sad part is that so many otherwise good people gave themselves selflessly to this process of destruction.

2. *Science* — The Western empirical scientific method was seen by those who practiced it as being neutral, objective, and devoted to the betterment of humanity. Unfortunately, those who practiced it really did not allow for any other valid system of producing testable, predictable results. Consequently they had no interest in ethnoscience, because it was not presented as science, in the Western sense, and was associated with a ritualistic, nonscientific cultural context. In other words Western scientists paid it no regard because they could not imagine it contained anything of value. After all, the land-use systems in the tropics were thought to be the result of chance, since the societies had not freed themselves from traditional "irrational" explanations of cause and effect. If the scientists had paused for a moment, they would have realized that 10,000 years of survival in a place represents quite an empirical record. Now it is recognized that ethnoscience — however it got there — produced valid results in many fields, and scientists are now searching through the shattered remnants of societies for ancient natural cures, farming methods, and other ethnoscience on the verge of extinction. In the context of science, mention has to be made of the arrival of Charles Darwin on the scene in the mid-1800s. His *Origin of the Species by means of Natural Selection* (1849), outlining as it did a theory of evolution, natural selection, and the survival of the fittest, gave new meaning to all those culturally superior terms that the Victorians and their colonial ancestors had been using. It now appeared scientific to say that the "backwardness" of the tropics was due to the accelerated development of a superior race. This is an aspect of science that continues to bedevil the literature to this day and weaves a confusing web of cause and effect. The tropics, at the time of Darwin, had been so brutalized by the coming of the Europeans that most of the observed differences were culturally induced rather than resulting from genetics. If a continent is subject to endless slavery, tyrannical "wars of pacification", and so forth, it is unlikely to evolve quietly along the path to progress.

3. *Economics* — The colonial system was quite unashamedly unequal economically. Initially the tropics would supply luxury goods through trade, but fairly quickly the source of supply was taken over and the growers came to work for the Europeans. It was because of this inequality that workers, farmers, and others in the tropics were never allowed to sell their labor in an open market, so what they were paid was determined by those who controlled the system. In the case of slavery, they were paid nothing at all. Somehow we now take it for granted that farmers and workers in some parts of the world are paid less than those in others. This is how that difference came about. Under the colonial system, the **value added** (the extra value added by turning a raw material into a finished good, distributing it, and selling it) was retained in the home country and the colonial empires became

suppliers of raw materials. This meant that there was little to no opportunity for the colonial areas to diversify their economic base, to accumulate capital, or to retain any indigenous control over their economic future. Furthermore, almost everything of commercial value that was grown was exported, which is why so many former colonial countries in the tropics have their colonial capitals at the edge of the territory and not in the center, where many of the old capitals had been. The colonial system, in effect, established a monopoly over different parts of the tropics by "enclosing" it, much as the setters had "enclosed" the Scottish highlands or the American plains. Thus the tropics was incorporated into a world system as a dependency of a power base elsewhere. That was to shape the distorted pattern of change that followed.

One interesting aspect of the colonial clash of cultures concerns the concept of "poverty". In this connection Professor Lynton K. Caldwell has written:

Poverty as now defined in the U.S. and the West, generally, is culture-bound. To Americans poverty is characterized by deprivation of material goods — homelessness and hunger are factors. But in many, so-called "poor" traditional societies people have homes, enough to eat, and relative security, yet we regard them as poor because they don't generate much in the way of (monetized) gross national product. Before commerce and Christianity intruded into many of these societies there is no evidence that the people thought of themselves as "poor".[3]

REFERENCES

1. Steuart, Sir James. *An Enquiry into the Principles of Political Œconomy: Being an Essay on the Science of Domestic Policy in Free Nations* (London: A. Millar and T. Cadell, 1757, reprinted by Chicago University Press in 1966).
2. Drummond, H. *Tropical Africa* (London: Hodder and Stoughton, 1889).
3. Caldwell, L. Personal communication (1992).

OTHER USEFUL READING

- Jacks, G. V. *The Rape of the Earth* (London: Faber and Faber Ltd., 1939).
- Kiernan, V. G. *The Lords of Humankind: Black Men, Yellow Men and White Men in an Age of Empire* (Boston: Little Brown & Company, 1969).
- Passmore, J. *Man's Responsibility for Nature: Ecological Problems and Western Traditions* (New York: Charles Scribner's Sons, 1974).

CHAPTER 9

Colonial Agriculture in the 20th Century

General Trends

Much of what happened in terms of the general direction of colonial agriculture in the early years of this century was a result of the increasing penetration of, and interest in, Africa. Although Europeans had maintained trading stations around the coast of Africa for centuries, the penetration of the continent did not begin in earnest until the end of the 19th century. Africa had a large peasant population and, except in such places as Algeria and Rhodesia (Zimbabwe/Zambia), the policy was to develop native agriculture rather than to alienate vast areas of land to European settlement. Up to this point the colonial interest in tropical agriculture had largely been taken care of by the community of white planters, with the colonial government playing a minimal role, such as establishing new land laws and keeping law and order. The role of the planter had been described as "bringing" skills and capital to create wealth, which was then distributed among the local population in the form of wages, food, and some rudimentary social services. The planters, it was felt, could generally look after themselves and so the government's role in tropical agriculture was essentially to provide the infrastructure for commerce and trade.

In Africa the colonial authorities had to deal with a new problem, that of commercializing a peasant population. The powers in Paris, London, Brussels, and so on were anxious that, at the very least, these territories should pay their way and that the grants-in-aid should be terminated at the earliest possible date. Another difference the authorities faced was that vast areas of Africa were suitable mainly for annual plants rather than for the humid, tropical tree plants or bushes such as rubber, tea, cocoa, and so forth, which were grown mostly on plantations. So the guiding interest of the colonial agricultural authorities became redirected toward the small, indigenous farmers and the need to bring them, as soon as possible, into the money economy. In doing this, the authorities were drawing these peasant farmers out of the tradition of subsistence and into the world economy so that they might now be affected not only by changes in the weather or plagues of locusts, but by decisions made on the commodity markets in London, New York, or some other foreign locality totally beyond their control. Essentially, the agricultural authorities saw themselves as having, at the most basic level, a revenue-raising function, in that they were involved in turning peasant agriculture into some form of commercial farming to serve the export market. In some cases, such as in Northern Rhodesia, where mining and urbanization developed rapidly, there was also a need to commercialize agriculture and to create a marketable surplus to feed a growing *local* nonagricultural population. Throughout the remainder of the colonial period, however, farmers were forced to concentrate on cash crops, thereby neglecting the food, or staple crops. This was to lead to great problems for countries in the tropics as they gained their independence.

To a large extent the general feeling, on the part of the colonial authorities, was that the food crops could take care of themselves. Indeed some territories such as British Guiana (now Guyana) actually issued an edict forbidding the cultivation of rice by the local population because it was felt that time taken growing that crop would draw labor away from the plantations. As a result British Guiana became a net importer of basic foodstuffs. Across the globe in Malaya, much the same attitude prevailed, and by 1940 the concentration of attention on rubber meant that only 34% of that territory's rice needs were home-grown; the rest was imported. However, this indifference was severely shaken by the incidence of famine, starvation, and diseases, which became fatal among such a weakened population. The French experienced such a famine in their Niger territory in the winter of 1925, the British in Uganda in 1918, and in Bengal in 1943. The initial response was to bring in famine relief, such as the 50,000 tons of maize imported into Kenya in 1943, and, later, to encourage or insist upon the cultivation of famine reserve crops such as cassava (manioc) along swamp margins or in corners of the family plot. Nevertheless, scant attention was directed toward the actual improvement of traditional or staple crops since there was no commercial gain to be made from the money invested in such research.

Colonial Research into Agriculture in the Tropics

The beginnings of organized research into agricultural production in the tropics can be seen in the setting up of botanical gardens, as was done in Ceylon (now Sri Lanka) and Java by the British and Dutch, respectively. The original purpose of these gardens was to test the genetic potential of plants found in other parts of the tropics for commercial development. The first of these botanical gardens had been established by the Dutch at Cape Town in 1694. Indeed the entire American coffee industry developed from a Dutch initiative when one coffee plant was sent from Java, via Amsterdam, to the Dutch colony of Surinam in 1718. The British developed their famous clearinghouse for the development and dispersal of plants at Kew in 1760. Once the initial phase of discovering and breeding new crops passed (though it was still going strong in German East Africa [now Tanzania] in 1902 when sisal was imported and bred from the Yucatan in Mexico), the attention of the botanists became increasingly concerned with crop disease.

One of the great hazards associated with the plantation economy had been that it was essentially a **monoculture**, in that it concentrated exclusively on one crop. This increased the risk of the development and rapid distribution of disease through the crop. Indeed the coffee monoculture of Ceylon was almost totally wiped out by just such an event when it was struck by the virus *Hemileia vastatrix*. Problems also began to emerge in the sugar industry in the West Indies and Java, where constant cropping without relief for the soils and without the addition of other nutrients from outside led to a steady decline in the yields. Indeed one of the most onerous tasks of a slave in the West Indies was carrying head-loads of manure (sometimes imported from England) onto the fields. The planters now began to look to the scientists not as the researchers in their ivory towers, but instead as the possible salvation of commercial agriculture. Indeed they actually contributed funds through export taxes to maintain the botanical establishments in some territories.

By the 20th century research was being conducted on a much more systematic basis with the formal support of the colonial governments. Most of the research establishments were commodity based, concentrating on one commercial enterprise. Examples would be the Sugar Research Station in Barbados set up in 1886, the 1914 Ceylon Rubber Research Scheme, the Ceylon Tea Research Institute in 1925, and so on. The research phenomenon is considered in more detail in the following chapter.

The lack of attention to traditional farming systems, to traditional or to introduced staples, or to local concepts of agriculture was to prove a sad error. It was based on the principle, as noted earlier, that traditional agriculture was unscientific and uncommercial, so there was really nothing to learn from it. Shifting cultivation was frequently described as wasteful or primitive, and yet when it was discovered in Uganda that constant cotton cultivation was breaking down the crumb structure of the soil, the proposed solution of rotating elephant grass (*Pennisetum purpureum*) was actually nothing but an adaptation of exactly those principles that had been used by the shifting cultivators.

The Organization of Agricultural Services

The 20th century has been marked by a growing direct involvement of government in agriculture both through the building up of support services, taxes, and levies, and through direct investment in production schemes. Agricultural officers are so much part of the tropical scene these days that it is difficult to imagine how recent a creation they are. The need for these officers arose mainly because of the special problems of the commercialization of peasant agriculture. The colonial departments of agriculture were nearly all created during this century: that of Southern Rhodesia (now Zimbabwe) in 1903; Uganda in 1900; Malaya in 1904; British Guiana, Trinidad, and Jamaica in 1908. The Colonial Agricultural Service was created in Britain in 1935. The role of the agricultural officer was described in the Uganda handbook as follows: "to investigate native methods of agriculture and to discover what is useful in them; to stimulate the improvement of the indigenous methods of cultivation or the adoption of new methods; to give advice to owners; to instruct native subordinate staffs; to supervise Government experimental stations and generally to assist in the work of the Department." Very often, however, it was difficult for an expatriate officer to get inside the systems being practiced on peasant farms and the tendency was, instead, for the officer to become an instrument for disseminating the results of research carried out on the experimental stations. This is not surprising since agricultural officers were technical specialists; they could hardly be expected to become amateur anthropologists as well. Another problem was that many expatriates were familiar with what constituted agricultural systems in the West, and they tended to work toward transforming the indigenous farming models in this direction. One agricultural officer of the 1940s noted that the "problem remains of trying to get the native cultivator to understand the benefits of the changes we have to offer." There was also a need on the part of the officers to understand the perspective from which the native cultivator perceived agriculture and crop production. Some of these problems are reviewed at the end of this chapter.

The governments, as we have noted, came to intervene directly in agriculture. More and more government money was being expended on support services such as roads, stock routes, veterinary attention, and so forth. But with the establishment of colonial development funds, money became available for intervening directly in production. This was generally on a large scale and is typified by the Gezira scheme in the Sudan. There the irrigation of vast tracts of land between the two Niles involved an expenditure that was inconceivable to the peasant farmer. The infrastructure was established by the authorities, and farmers became tenants under a parastatal structure that many countries regarded as a model for *transforming* agriculture (i.e., commercializing it instantly) rather than for steady improvement by exhorting the peasantry to change. Other schemes were developed by the French in Mali and along the Senegal River. The scheme in Mali (the *Office du Niger*) failed because the price of the main crop, rice, could never match the expense of maintaining the scheme. In the Sudan the high price of cotton made Gezira a model success story for many years. To fund many of these developments, the governments imposed export taxes on several major cash crops, which built up reserves of money that were used both for stabilizing the price paid to the small grower and for building roads and other infrastructural developments.

Technological Change

Many of the European farmers who came as settlers had little or no experience of cultivation in the tropics (or anywhere else in some cases!). Consequently they introduced methods and implements that caused serious damage. The exhaustion of the soil by European maize monoculture in parts of Kenya is well known. Experience gained the hard way led to a steady improvement in methods, but the sudden introduction of new techniques into peasant farming often had a disastrous consequence. The deeper metal plough put to work on the rice fields of Southeast Asia tended to penetrate the impermeable clay layer and cause the drainage of water through the soil; the introduction of tubewells in the Punjab led to horrific losses of agricultural land through salination resulting from over watering and poor drainage. The random introduction of boreholes in the semi-arid grazing areas led to a concentration of animals and to circular areas of erosion and destruction on a wide scale. Sometimes the authorities themselves applied modern methods out of context such as the almost legendary groundnut scheme initiated in Tanzania just after World War II. This was precipitately developed to produce vegetable oil for a war-torn economy in Europe using leftover heavy machinery from the military. The lack of research and the unsuitability of the methods (e.g., leaving the ground bare and exposed to the erosive action of wind and water) resulted in losses variously put at between $20 and $80 million.

Some Problems
Prices

We have already noted the disease risk associated with monoculture; additionally, the peasant drawn into the money economy was at the mercy of price changes on the international market. The planters had suffered from this for a long time. Indeed, in the 17th century, there had been a halving of the price of tobacco as a result of overproduction. Cinchona — the bark of the South American *cinchona* tree, which was the source of quinine — suffered the same fate later and there was the extraordinary collapse of rubber prices from 12 shillings per pound in 1910 to a mere 2 pennies per pound (one seventy-second part of the 1910 price!) in 1921. The peasant cocoa farmers of the Gold Coast (now Ghana) actually rioted in the 1930s over a sudden fall in the price of their product. Colonial governments attempted to handle this situation by establishing price stabilization funds; with this, a long tradition of government intervention and control in agricultural pricing began. This soon became another form of taxation on the rural population.

Land Tenure

Associated with the breakdown of land was the question of land tenure. In Africa, attention was focused on the conflict between the requirements of good land management and the customary tenure of land; that is, it was common land that was, at the same time, everyone's and no one's. This was perceived as a particular menace to the semiarid grazing lands, where stock populations were thought to be rising at an alarming rate as a result of veterinary and water development programs. Nor did anyone take responsibility for controlling the grazing. In parts of Africa another conflict was seen: that between the perennial nature of many cash crops, such as coffee, and the lack of any secure land tenure system (meaning, in general, "ownership"). Gradually the nature of land tenure was changing to a de facto ownership where land was being bought and sold. However, this was not codified under customary law, and small-scale farmers were thought to have inadequate protection for their investment of capital and labor. On the other hand, the authorities were extremely worried about what was happening in India, which had a well-developed land market. Peasants were falling into debt wholesale and a powerful community of absentee landlords was acquiring the land and extracting wealth from it through sharecropping arrangements. The peasants of Zanzibar had been almost totally beggared by debt and had sold off their land to Indian merchants, who failed to maintain the clove crop adequately. The government was eventually forced to buy the land back.

In conclusion, the colonial economy brought the peasant farmer into the wider world. The greater part of the profits from the cash crops (the "value added") accrued in the metropolitan countries where the crops were processed, brand-named, and sold. The plantation economies were shaken violently by price fluctuations, and smaller, private plantations came to be replaced by foreign companies, as happened with the direct investment of American money into the banana plantations of Central America. But the preoccupation with cash crops and export earnings would lead to a monumental problem for the newly emerging independent countries of the tropics: the breakdown of food supplies.

USEFUL READING
- Masefied, G. B. *Handbook of Tropical Agriculture* (Oxford: Oxford University Press, 1951).

CHAPTER 10

The Development Dilemma

"Human beings have broken out of the circle of life, driven not by biological need, but by the social organization which they have devised to 'conquer' nature: means of gaining wealth which conflict with those which govern nature."[1]

Evolution from Colonialism . . . or More of the Same?

We saw in earlier chapters how the long-evolved traditional land-use systems were unequally incorporated into the world economy during the phase of European expansion into the tropics. However, that phase is over, at least formally, except for a few, usually minor exceptions. In theory, starting with the liberation of Latin America in the 1820s and the rest of the tropical world between 1947 and the 1980s, we should be in the third phase, during which priorities are again determined by those who come from and live in the tropics according to their own traditions, needs, and principles. Unfortunately, it does not seem to have worked out that way.

The leader and trendsetter in this field of newly independent countries was India, which achieved full self-government in August 1947, and partly as a consequence of its considerable size, was perceived as drawing up the blueprint for the transition from colonialism to independence. Its thinking was strongly influenced by what had happened in the Soviet Union after the revolution of 1917, though India never espoused the totalitarian Marxist philosophy that Stalin used to achieve his dreams. The Soviet Union was one of the only examples of a large country that had deliberately set about changing its social order and reordering its economic priorities: the very challenges India perceived for itself. In the USSR, however, the agricultural sector had been quite deliberately made subservient to the drive for industrialization: a "transformation" approach to development. In short, all the steps of Western development since 1760 were to be leapfrogged in a rush to industrialization. Agriculture would be taxed to achieve this, and the benefits were to somehow "trickle down" to the countryside later. There was, in fact, no real thought given as to what *sort* of society and economy was to be built. Essentially, the model was one of "modernization", by which was meant "catching up" with the West. The bill for this was to be paid by the farmer.

In the West the visible signs of agricultural modernization were the elements of capital scattered through the countryside: the tractors, combines, buildings, fertilizer stocks, pumps, and so on. Since those who lived in the tropics perceived the colonial epoch as a time of asset stripping, the logical interpretation of "catching up" in this context would be the *accelerated acquisition of capital*. Thus began the ill-fated mass transfer of Western technology and money into an environment lacking trained mechanics, spare parts, credit, and almost everything else necessary to make capital work. The capital-intensive transformation of tropical agriculture was a disaster, starting with the Tanzanian Groundnut Scheme in which Britain lost millions pretending that East Africa was Wiltshire, continuing with the dispatch of Soviet snow plows to Guinea, and ending with the appalling state farms of Zambia.

Why the Transformation Model?

The reason why newly independent governments in the tropics turned their backs on their own farmers and their own history can be explained in terms of several perceived priorities.

The Need to Catch Up

This was alluded to earlier, and resulted from the fact that the decision makers in tropical countries had often been educated abroad or in an imported context at home. This training deemphasized indigenous practices and values to the point that, in Uganda, students who misbehaved were punished by being put to work on the school farm! This "catch-up" mentality conveyed the idea that there was a universal model of development and that countries were further back, or further along, the same path. The archetypical expression of this orthodoxy was W. W. Rostow's 1966 classic, *Stages of Economic Growth*, in which each country could plot its "progress" along a normative "S" curve through "take-off" to high mass consumption.

Given such a linear, universalist perspective, local decision makers are just as unlikely to look into their own society for answers as were their colonial masters. Once again, indigenous conditions tend to be viewed as "obstacles" on the path to a commonly held view of progress. The only diversion along the way would be to shop at one of the stores selling foreign ideologies, so the choice would be between 19th-century Western free-market industrialization or some brand of 19th-century social engineering and command structure. While none of this had anything to do with Bolivia, Zaïre, or Papua New Guinea, it cost a lot of countries a great deal of time and hardship. Latin

America struggled forward with a totally fossilized, centuries-old Iberian model of the worst form of semifeudal exploitation. Africa experimented with socialistic mind games and posturing, which bankrupted the entire continent, while Asia was divided among mind-numbing bureaucracy, ideological hot wars, and some of the most aggressively successful capitalism the world has yet seen, often achieved at terrible environmental cost.

The mindset that accompanied the transformational imperative was well displayed at the United Nations Stockholm conference on the environment in 1972, when the delegates of the developed countries were urging all nations to become more conscious of the need for environmental stewardship. Delegates from developing countries observed then that this was the perspective that comes from wealth and industrialization — the same wealth that gave a country the resources to do something about pollution. The developing world was still seeking to build the basis of industrial wealth and would look to the cost later. For now, their environmental problems were part of a larger issue of poverty. They were not going to give up the race now that they had just entered it.

The Rejection of History

Along with their colonial past, most of the newly independent countries chose to consign their *entire* past to what Trotsky called the dustbin of history. New nations with no indigenous provenance had been created, and so the surviving realities of the past tended to disturb the process of building artificial nations and to encourage fragmentary tendencies. Tradition was baggage on the road to modernization and was best dumped to lighten the load. Often the only common language of most newly independent nations was the language of the former colonial power. Hence, the former Spanish colony of Equatorial Guinea remained blissfully isolated as the *only* Spanish-speaking state in Africa at independence.

Independence gave these countries a unique chance to reevaluate what had happened to them, but sadly the opportunity was totally lost in most cases. Here and there a light glimmered: Nyerere's attempt to create in Tanzania a blend of traditional and modern in his concept of *Ujamaa* socialism; Gandhi's emphasis on village life. The former was destroyed by a totally unsympathetic bureaucracy, the second by an assassin's gun and a succession of urbane industrializers.

The Pressures of International Assistance

As most of Africa and Asia became independent after 1947, an enormous bilateral and multilateral system of support was constructed to accelerate the transformation of these new states and belatedly those of Latin America, too. Best known among these, the International Bank for Reconstruction and Development (or "World Bank"), was in fact initially focused on the *reconstruction* of war-torn Europe in an attempt to halt the westward flow of communism. This was part of the Marshall Plan, named for General George Marshall of the

United States. The plan was extremely successful in replacing Europe's industrial and agricultural capital, which had been destroyed in five years of bitter fighting. The highly visible German "miracle" was, for the West, the equivalent of what Stalin's transformation of the U.S.S.R. had been for Indian planners. At the same time, it was an alternative to the communist principles that underlay the Stalin model. Few stopped to think that both the German and Japanese "miracles" had been miracles of *reconstruction* in already "developed" countries, replacing machines, factories, railroads, etc., not human capital. This simply was not the context in tropical, developing countries.

An entire new industry emerged, engaged in transferring Western capital and technology to aid the "transformation" of poor countries. This was not dissimilar to, and was certainly as arrogantly self-confident as, the thinking that had prevailed when the Europeans first poured into the tropics: "We come bearing the answers to all your problems, which, coincidentally, were the answers to ours as well." Parallel to the Western aid effort was a similarly Cold War-inspired effort by the centrally planned ("communist") states to ship advice, specialists, and capital goods. The tropics became a playground for the ideologues: Cooperatives or communes? Peasants or yeoman farmers? State farms or entrepreneurs? Unfortunately this contest had the same passing contempt for indigenous knowledge and culture as had the now passé colonialism.

This aid relationship is characterized by several features that, for decades, had a decidedly negative effect on the tropical environment.

- Foreign aid concentrates almost entirely on the foreign currency capital gap. That is, it is there to provide the objects and technical advice that are available only for hard currency on the world market. These commodities are usually in short supply in developing countries. Consequently, what comes through the aid network is almost entirely from, and designed for, the temperate, capital-intensive regions. All technology has a context, which travels with it when it is transferred.
- Aid is generally *project based* and tends to focus, at best, on the internal considerations of the individual project. Donors and lenders relied on the government of the host country to construct broader environmental provisions in their planning and development strategies. Normally these possibilities have been absent, and so the project provides the total horizon for environmental thinking. Since most aid is connected with projects (providing grants or loans for the achievement of a specific activity in a specific time), there is a powerful inherent tendency to scale up. Larger projects carry proportionally lower overheads in their design and implementation stage. It is easier to package a major dam or a plantation project than it is to do something meaningful for hosts of small farmers. And yet most of those who live from the land in the tropics do so at a minuscule scale, well below 5 ac.
- Aid is supposed to fill a gap in the needs of developing countries *as perceived by those countries*. Thus it can only be as good as the recipient government, since it is a political, government-to-government relationship. In other words, the recipient countries are, in theory, supposed to identify

priorities, and in line with these priorities, the projects that are economically best suited to meet needs. In reality a sort of circular reasoning prevails in which the countries have come to identify the sorts of things that donors and lenders "like to do". Projects are generated according to that priority rather than indigenous priorities. Since most aid is in the form of loans that must be repaid, there has to be a foreign currency-generating component somewhere in the "stream of benefits" to ensure this happens. This is frequently not congruent with food security or other subsistence needs.

- Aid is supposed to be "economically viable", which means that it ultimately pays for itself (this is with the exception of relief aid, which is usually targeted at dealing with some specific catastrophe). The definition of "economically viable" is, however, critical, and it is normally the prerogative of the donor or lender to define the parameters, the methodology, and the threshold limits. All these points tend to drive aid into perpetuating both the external definition of priorities and its own severely constrained view of what is "economic". Most aid accounting up to now has given almost no recognition to the cost of environmental damage, the ultimate loss of natural resources, or the intergenerational costs of "development".

Growth or Development?

During the 1960s there was a real crisis in the thinking behind aid from the West. Primarily, foreign aid was designed to build a secure free-market base to counter the threat of world revolution inherent in communism. Even though the national economic statistics of most developing countries grew impressively through the decade, the scale and intensity of poverty *also* grew, providing recruiting grounds for insurgency in all three "southern" continents. The paradox resulted in a new approach, encapsulated in Hollis Chenery's World Bank publication *Redistribution with Growth*.[2] This book examined the fact that getting richer on a national scale does not mean that everyone becomes better off. And so aid became "targeted" at certain sectors and social groups (agriculture, rural development, the "poorest of the poor"). Again this targeting could be only as successful as the policies of the recipient governments through which aid *must* flow. Sovereignty often negated the various creative fads and fashions that worked their way through the aid industry.

The main concern was the *equity* argument: Who benefits from the growth? There was really no doubting the *growth* theology itself. Production remained sacrosanct; distribution was at fault. None of this touched the environment, except for the fact that since the poorest farmers had been kept securely out of the economic system, they were not given the means to damage the environment beyond the sheer buildup of their numbers. Most of the "targeting" was, however, just an exercise in creative vocabulary, and there was little substantial change in what was happening. At the root of the problem was the fact that most developing countries were getting further and further into a trap, which limited their options and increased the destructive cost heaped on the environment. The elements of this trap were, in addition to the modernization/socialist reconstruction paradigms and aid distortions, as follows:

- Well into the 1970s most of the aid was capital intensive and depended on the large-scale introduction of transformational machinery (tractors, pumps, fertilizers, etc.). With the 1200% increase in oil prices between 1974 and 1980, the economics of this transformation were grossly distorted against the tropical countries (most of which do not have their own oil resources). This led to a dramatic need to obtain *more* foreign exchange to pay for the imported oil to run fixed investments. The only way to get that foreign exchange, in most cases, was to *reinforce* the colonial economy by exporting raw or semiprocessed primary products. This worked dramatically against innovation and change and encouraged asset stripping.
- During the boom time of the 1960s, countries had no trouble borrowing money for "development" projects and for covering shortfalls in government spending on consumption and services. In the 1970s, as energy prices took off, banks were, curiously, still falling over themselves to lend money to poor countries. The reasons for this phenomenon were threefold: (1) the banks were awash with petrodollars from oil-exporting countries that had limited investment potential at home, (2) lending to countries was considered a "good risk" since countries, unlike people, would surely not default (which was the ultimate proof that bankers simply never learn — countries have defaulted on national loans continually since the Middle Ages), and (3) everyone thought that the crisis was a purely temporary one and that countries would absorb the shock and recover.

In 1982 the mythology collapsed as Mexico declared itself unable to meet the interest payments on its accumulated debt. Suddenly there was no more credit, there was a huge threat of foreclosure and collapse, and the financial markets of the world went into a tailspin. In the developing countries this meant that the *real* priority now was to find hard currency to meet annual debt repayments: a situation that became worse monthly as the value of the U.S. dollar (in which most debt is calculated) went skyward with Reagan-era borrowing and as interest rates took off (also due to Reagan-government borrowing). The visible horizon of most developing country governments was now reduced to a year and was entirely dominated by debt-handling priorities. To the pressure for growth had now to be added the pressure to repay — and the weight of most of this was going to fall on the natural environment. This pressure to use natural resource exports to pay for short-term obligations reinforced the "colonial" economy, especially in terms of the long-standing colonial disregard for traditional food staples and their obsessive preoccupation with cash crops. The colonial pattern was further emphasized by the barriers erected by the developed world to the "value-added" component of primary exports. Unprocessed raw materials could freely enter the developed countries, but once some value was added to them, turning wood into furniture for instance, then the tariff barriers immediately posed a problem, restricting the ability of countries to export fewer products of a higher unit value. In effect the newly independent countries escaped from being exploited by others only to end up exploiting themselves for the benefit of the same masters now recast in a new guise. *Plus ça change . . . .*

There emerged during the 1980s the interesting paradox of, on the one hand, increasing awareness of the urgent need to protect and conserve the tropical environments and, on the other, the urgent necessity to mine the same environments to the maximum to pay the bills. The net capital flow has, for decades, been *out* of the tropics. In the political arena, conservation (or sustainability) and growth were seen as mutually incompatible and conflicting. Environmental impact statements for large-scale development projects were perceived of as "costly" and as "time-consuming delays" impeding necessary progress and growth. The longer-term perspective of the World Bank was rapidly displaced by the "balance of payments" and "standby credit" annual priorities of the International Monetary Fund. One day governments in tropical countries were besieged by conservationists from the north berating them for their mishandling of forests, rivers, soil, water, and almost everything else the next day at home they were confronted with demands to meet unfulfilled promises, as well as by the cold, hard figures from the men in suits from Washington. Politics, anyway, has an innate tendency to seek answers to the question, "What have you done for me *lately*?"

Developing countries found themselves increasingly in a dilemma that took the following form. Rural poverty encouraged mass migration from the countryside. Unlike in Western society in the 19th century, on this occasion there was nothing waiting for the migrant at the end of the journey. To keep the lid on the potentially explosive situation of burgeoning cities full of poor and hopeless people, the governments subsidized basic foodstuffs, which is another way of saying they "taxed" the countryside. This tax led to systemic, ongoing poverty in the rural areas and the incentive for even more people to leave for the cities. It also led to the massive buildup of rural poverty, hopelessness, and the consequent ravaging of the natural environment by people who see no alternatives, and who have a somewhat natural, short-term desire to survive. This endemic poverty is the root of the tropical environmental malaise.

In Africa the crisis was particularly acute since the economies of most of the countries on that continent were actually going *backwards*: per capita income and food consumption were *dropping* throughout most of the 1970s and 1980s. Eventually, apocalyptic scenes such as those in the Sahel in the period 1968 to 1972 and those in Ethiopia in the mid-1980s provided the crises that seemed necessary to trigger some form of sea change. Whatever was being done was clearly not working. Furthermore, it became increasingly difficult to perceive what was happening as a natural disaster, though it was generally portrayed as a drought or famine by the authorities. To some extent the crisis was the result of what happens when we imagine that bureaucracies can be entrepreneurial. The people had become powerless victims of a grotesque state apparatus playing games of social engineering or protecting semifeudal privileges.

In the West, the general awareness of the environmental failures of its own economic system and the mortgaging of the future to short-term gain helped create a climate for a "new economics" that could bring the environment back into the cost/benefit equation. New instruments, institutions, and actors appeared on the legislative/administrative scene. Most project design, through which "development" happens in many tropical countries, continued to concentrate on what might be called a financial return basis: the costs of the inputs measured against the flow of benefits — usually calibrated in monetary terms because debts have to be repaid. There was sufficient creative juggling of imputed social costs and benefits to the extent that this author was once asked by an aid mission, prior to the design phase of a project, "What rate of return should we settle on?" The methodology was, apparently, capable of working back from any number. There was, however, no place for a consideration of environmental costs and benefits. They were, as they had been in the West until the late 1960s, *externalized*, to be borne by the community at large and by succeeding generations in particular. So, when clearing an area of forest for a dam, as an example, certain questions failed to appear in the analysis, such as:

- What will be the cost of replacing the inherent soil fertility once the forest is gone?
- What would have been the economic *alternative* of sustainable use of the forest for a range of renewable resources?
- What role did this area play in the traditional economy and cosmology of the indigenous population?

The conventional questions asked, instead, would be: How much is it going to cost to clear this lot? What will we get for the trees? What is the "economic" return on the changed activity? The last question asks how much profit, in a narrow monetary sense, can be made after debts are paid, on a year-by-year basis.

In 1991 the World Resources Institute (WRI) in the United States reworked the national accounts of Costa Rica to account for the environmental losses incurred by conventional short-term growth chasing. The conventional picture tells us that Costa Rica grew by 4.6% annually between 1970 and 1989, which is impressive. If, however, we put a value on the forests, soil, and fishery stocks, one quarter of this growth disappears immediately. Forestry alone accounts for 85% of the depreciation in that country's natural resources. Alarmingly, annual depreciation was 70% greater in the last 6 years of the study period than in the previous 12. In short, the WRI is saying that it is totally misleading to treat resource *depletion* as current income and yet that is what is happening. They propose that the standard framework for national accounts, which supposedly measure development, is changed to correct the current view that makes sustainability "unprofitable".[3]

Pressures on the international institutions involved in lending and donating funds to developing countries and on the governments of both lenders and borrowers, along with the heightened public awareness through international "lobby groups", have rendered the international externalizing of environmental costs as untenable internationally as was the same phenomenon nationally in, say, the United States. There is now a fairly clear distinction between "economics" and "ecology", as summarized in Table 10.1.

Table 10.1. Ecology and Economics: Some Comparisons

Ecology	Conventional "Western" Economics
Based on the laws of natural science: objective/neutral	Subjectively based on prevailing scarcity/social values
Quantifiable — high probability of predictability	Quantifiable — lower probability based on aggregate behavior patterns
Medium of measurement — an absolute	Medium of measurement — a surrogate
Stresses *limits*, balance, and stability	Stresses *growth*, imbalance, and dynamic qualities
Stresses *process*: energy moving through a system	Stresses *process* of exchange in the marketplace
System-based (ecosystem)	System-based (exchange system)*
Ultimately self-regulating (balance, entropy, thermostat concept)	Self-regulating through the hidden-hand concept of scarcity and price regulating the market
Holistic	Mastery concept of Judeo-Christian tradition in the West. Attempts at more pantheistic "Zen" models have been made, such as that of Schumacher
Operates independent of the human value system, though its health ultimately depends on the value placed on its sustainability by humans as long as they are around. After that nature can look after itself	Expresses the value system at the time and may or may not include some imputed value for the survival of nature in a sustainable form
Time scale not necessarily related to human life span	Hard to push perspective beyond the human life span or government term of office
Natural resources as capital that has to be managed	Natural resources sometimes regarded as free good with little regard to the short-term depletion of capital
Long-term perspective	Short-term has been conventional, but that can change

*This reflects "free-market" economics. Marxist economics believes in the ownership and control of resources by the state, in which decisions are made by the command structure, supposedly for the greater good of the people.

This contradiction between short-term economics and ecological sustainability had little to do with the chosen ideology of any government in the developing world. Marxist economics puts emphasis only on the value given anything by the labor input. It gives no intrinsic value to nature or resources per se. Attractive though the idea was to many ideologues in the developing world in the 1960s, the state, as steward of nature within some socialist framework, turned out to be as destructive and rapacious as any caricature capitalist.

In the West, change in the value placed on environmental conservation, protection, and sustainability has been largely brought about through the intervention of public opinion in the legislative process. This activity has usually resulted from powerful lobby groups and environmental interests. In the United States, for instance, no political representative these days can afford to be without an "environmental position". In developing countries, such

Text Box 10.1
The Costs and Benefits of Development in Central America

"The Central American states undertook ambitious development programs in the 1960s and 1970s to raise their standard of living and bolster economic independence. Though there were marked differences from country to country, they all embarked on development strategies based on agricultural exports and industrial development to achieve import substitution. Only a handful of industries has succeeded, and the debt undertaken in the process is today the single largest impediment to growth. The production of export crops — principally coffee, bananas, sugar, cotton and cattle — was modernized with large investments in mechanization, pesticides and fertilizers. Throughout the region, farmers who rented or farmed small plots without title were divested of their farms to make room for the export crops. These export crops accounted for one-half to two-thirds of export earnings in El-Salvador [sic], Honduras and Guatemala, but intensive monocultures have degraded the resource base, and the shift in land use meant a decline in food production. In the Costa Rican lowlands, for instance, small-scale agriculture virtually disappeared before the banana plantations in the 1960s, while corn production declined 87 percent in 6 years. In Nicaragua in the same years, it was cotton, whose huge sales to Japan made it the leading export.

"Each of these crops had its environmental costs, but it is deforestation for cattle ranching that has caused the most destruction. From 1960 to 1980, Central America's forests declined from 60 percent to 40 percent of the territory — with two-thirds of its cultivated land in livestock. Since 1963 the World Bank has lent funds for cattle ranching to every Central American country except El-Salvador [sic], and has provided more credit for livestock than any other kind of agricultural activity. The Inter-American Development Bank (IDB) also recognized cattle production as "highly suited" to the region and a promising earner of much-needed foreign exchange. . . .

"Today, the highest rate of deforestation in the hemisphere is found in Central America. Fuelwood is already in short supply, and some countries will be forced to import lumber in a few years. Dams and rivers silt up faster than expected as soil is carried off. . . .

"The development strategy of the past generation has left the Central American states more dependent than ever on the United States and the rest of the industrialized world."

Source: Brown, J. W. *Poverty and Environmental Degradation* (Baltimore: WRI Publications, 1988) pp. 8–10.

lobbies — which are, worldwide, often a middle-class domain — are poorly developed, and those with political influence are often heavily involved in benefitting from the economic status quo. This conveys to some outsiders the idea that people in developing countries are somehow unaware of or don't care about the destruction of the environment. There were, it should be remembered, virtually no environmental movements in the London of Charles Dickens. That was not because people didn't care about living in almost unbelievable filth and

---

Text Box 10.2
Reasons for Questioning the Benefits of Conventional Economic Growth in the Tropics

- By 1980, 25 to 40% of all original tropical forests had been lost.
- 50,000 mi^2 of land is lost to forest clearance each year.
- 20% of all irrigated land is classified as waterlogged or salinated.
- 15 million acres each year are added to the list of land moderately or seriously desertified.
- Without changes in conservation, total rain-fed cropland in the developing world will shrink by 18% by the end of this century, with a 29% drop in productivity.
- Two fifths of Africa's nondesert land is at risk of desertification.
- Flood-prone areas, as a result of upstream deforestation, increased from 50 million acres in 1970 to 100 million in 1980.
- 10,000 persons now die annually from poisoning as a result of uncontrolled pesticide use.
- Of 33 countries in the tropics now exporting wood, 23 will be net importers by the end of the century.
- Pesticide resistent insect species grew from 25 in 1974 to 432 in 1980.
- By the end of this century, another billion people will have been added to the world population (based on 1988 figures). More than 90% of these people will live in the developing world.

---

squalor. It was because they were *powerless* to do anything about it. They were working 80-hr weeks, leaving little time to eat and sleep, and they had no vote anyway. If that sounds like many developing countries, then the resemblance may be more than coincidental. There are barely any democratic governments in Africa, there is gross distortion in access to natural resources in Latin America, and a large percentage of Indian farmers are in thrall of intergenerational debt. These are not prime conditions for the development of active lobby groups.

Throughout most of the postwar period, we have lived through times in which ecology and conventional Western economics have been in conflict in the tropics — a conflict that was sharpened by debt and the oil crisis. The environment is mortgaged through the exigencies of poverty and hopelessness, rather than through the calculated onslaught of malevolent capitalists. Many of the trends we observe in all three southern continents are continuations of colonial miscalculations: the emphasis on cash crops over food crops, the essentially imported models of "development," the structural preference shown to the urban rather than the rural sector, the use of the rural sector as a tax base for modernization and the extraction of capital, the transformational approach rather than improvement of what is there, and a failure to include the environmental costs and benefits into either national policy or project-appraisal techniques. The future of environmental sustainability in the tropics depends on breaking this mold once and for all.

REFERENCES

1. Commoner, B. *The Closing Circle* (1972), p. 299.
2. Chenery, H. B. *Redistribution with Growth* (New York: Oxford University Press, 1974).
3. World Resources Institute. *Accounts Overdue: Natural Resource Depreciation in Costa Rica* (Baltimore: WRI Publications, 1991).

OTHER USEFUL READING

- Bartelmus, P. *Environment and Development* (Winchester, England: Allen and Unwin, 1986).
- Blaikie, P. *The Political Economy of Soil Erosion in Developing Countries* (London: Longman, 1985).
- Blaikie, P. and H. Brookfield. *Land Degradation and Society* (New York: Methuen, University Paperbacks, 1987).
- Carpenter, R. A. and J. A. Dixon. "Ecology Meets Economics: A Guide to Sustainable Development," *Environment*. 27(5):6 (1985).
- Cottrell. A. *Environmental Economics* (London: Edward Arnold, 1978).
- Dasmann, R. F., J. P. Milton, and P. H. Freeman. *Ecological Principles for Economic Development* (London: John Wiley & Sons, 1973).
- Gupta, A. *Ecology and Development in the Third World* (London: Routledge, 1988).
- Redclift, M. *Development and the Environmental Crisis: Red or Green Alternatives?* (New York: Methuen, University Paperbacks, 1984).

CHAPTER 11

Changes in Research: Systems and Revolutions

Researching Research — An Endless Loop?

Sometimes the world, and especially the academic world, seems full of people "doing" research. Very rarely does that research encompass research itself as a valid subject for serious investigation. Perhaps the reason for this is that many seekers after truth look upon research as a tool box: the car mechanics of problem solving. In Chapter 13 of this book, reference is made to the assumed *neutrality, objectivity,* and *detachment* of science as a problem-solving methodology. This is, alas, far from the truth, and it behooves us to consider how we define and become aware of problems, how we rank them in terms of their appropriateness for study, how we allocate resources to the solution of these problems, and how we evaluate the success — or otherwise — of our efforts, *including* the problem-solving system itself.

Right away it looks as though we are entering some intellectual black hole by saying "If our approach to research is flawed, then how will we discover as much, since presumably we shall research even that question by the tried and true methods?" Not exactly. If the evidence accumulates that our responses to generally agreed problems are not working, then there has to be an intrinsic flaw in what we are doing. As we look around the globe at the moment, we may take comfort in the fact that there have never been so many institutions — literally thousands — researching all aspects of the physical environment and the ways we use it. Furthermore, there has never been such an armory of technological tools to combat the decline and degradation of the environment and to alter its chemical composition, physical qualities, the distribution of resources such as water, and so forth. Never in history has so much money been available for the support of serious investigation into the whole array of environmental problems.

And yet, the reality is that we are losing huge swathes of the globe to desertification, continuing with the sterilization and exhaustion of former tropical forest lands, and observing the declining health and well-being of both the seas, as well as the air we breathe. A true cynic, examining the record of the last 25 years, might be cruel enough to plot, for instance, the research effort in the field of desertification against the loss of productive land to that phenomenon. The relationship that appears on the graph would probably be sufficiently positive to lead to the speculation that research was "causing" the problem, since the two seem to grow in tandem! It is closer to the truth to say that research is either (1) not addressing the entire problem in terms of its causal relationships or (2) being incorrectly applied through public policy and private action.

In this regard there are two popular cries to be heard at international gatherings from those connected with major environmental problems. The first comes from the rich, developed nations when discussing such phenomena as acid rain (not yet a tropical problem, but the one that best exemplifies the attitude). When Canada accused the United States or when Sweden accused the United Kingdom of *causing* the problem, there was an instant response, "We need more research," meaning that the causality, or even the true existence of the problem, has not been proved. Sadly, proof in the scientific sense may be a long time coming for many of these problems, and it may arrive in that "long term" by when, as Galbraith assured us, we shall all be dead.

The second response to research is more commonly heard from the developing (mostly, tropical) countries at international conferences. Here the general frustrated cry from the political delegates is, "No more research. We have all the research we can handle. We want *action*." In this case the general feeling is that the research is of an inapplicable, stratospheric, and totally technical type offering policymakers little in the way of concrete, usable, and realistic programs of action. So, for one group of politicians, this is the required godsend to avoid action, and for the other, it is the obstacle to serious and immediate action.

Factors Shaping the Nature and Effectiveness of Research

Research has its own environment governing problem definition and what gets done. Briefly, as all researchers know very well, the following factors are influential in shaping the research agenda:

1. Problem definition. Someone, somewhere has to define the problem. We will see in the section on the administrative trap (Chapter 12) that what constitutes a problem may be defined not by natural circumstance, but by the terms of reference of a public organization that sees only what falls within its legal and administrative powers. A problem may not "exist" until someone in authority chooses to recognize it. (Think of pollution in the

United States, which did not become a serious "problem" until the legislators were forced to deal with it, or think of famine in Ethiopia, a problem that the Ethiopian government simply denied.) Not all truths are held to be self-evident — some have to be demonstrated to be so. There is a considerable element of subjectivity in terms of when a situation becomes worthy of research, and how that subject is defined may depend on the perspective of an individual or on the organizational perspective of institutions.

2. Research depends on resources, and the allocation of these resources is itself a subjective process further muddying the idyllic waters of detached scientific neutrality. There is a stage at which priorities have to be determined and a judgment made as to how much can be allocated to each activity. If a problem is embarrassing to a government, as were the deaths of those starving people in Ethiopia in the 1980s, then the government may simply deny the existence of the problem.

3. Knowledge has traditionally been organized (its epistemology) along Cartesian lines into disciplines or "subjects", and the funding structure, similarly fractured, has institutionalized — some would say "fossilized" — the ability to perceive broad, interactive problems. Unfortunately, most environmental problems are of this sort, requiring the melding of both the social and natural sciences for any real understanding to emerge. Researchers are familiar with being told that proposals for studies of land use are neither "natural science" nor "social science" by the respective funding bodies, and so this vital area of interactive study dies under a *real* dead hand of tradition: academic tradition.

4. Over 90% of research and development is carried out in the developed world. As far as the tropics are concerned, this is another complicating factor. A consequence of this is that the results of this activity have then to be *transferred* to the often totally different cultural environment of the poorer, tropical nations. In this case the cultural context that nurtured the research (labor scarcity/capital availability for instance) may be totally inverted relative to the recipient society. At least it is a dangerous assumption to hold this context as in any way *given*. A simple example of this fallacy would be the question of maintenance. Where is the support mechanism of trained mechanics, spare parts, etc. for imported high-technology agricultural equipment? The average life of a tractor in Zambia, for this very reason, is only 3 years.

5. There is considerable difficulty in reconciling the "hard" sciences and the "soft" sciences to working together without prejudice on serious issues. This is barely possible in the universities; it is even harder in the field. Each maintains a polarized caricature of the other: One group is seen as completely naïve about policy and behavior, the other as totally lacking in the rigid disciplines of proof and an appreciation of the inevitabilities of natural order.

Research in the Colonial Times

It is not necessary to treat the colonial era in great detail. However, things happened then that had a formative influence on what carried over after the independence of many African and Asian nations following World War II. The

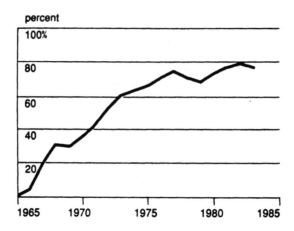

Figure 11.1.   Share of wheat land planted to high-yielding varieties in India, 1965 to 1983. (Reprinted from Brown, L. R. "Feeding 6 Billion," in *Environment 90/91*, Worldwatch Magazine, Worldwatch Institute, Washington, D.C., 20036, 1990. With permission.)

Latin Americans tended to take their lead from the patterns prevailing in the United States. Both instances provide serious continuity and transference problems.

As we have described earlier in this book, the Europeans who came to the tropics really did not believe there was any rationally derived scientific knowledge in the areas they came to govern. Science was, to the Europeans, exclusively a product of the Western Enlightenment. Thus the whole thrust of colonialism was *unidirectional*. There might have been some sense of enquiry about the physical environment and what it had to offer, but there was no parallel curiosity into the cultural systems that had exploited the environment for centuries. Research looked for the most effective ways to "modernize" these areas and bring them into the world of trade and commerce. Even anthropology came to be known as the "handmaiden of colonialism" because it was so often used to investigate how local people might better be convinced to change their old ways in favor of the colonial agenda.

To the confident Western mind, research was totally utilitarian in a narrow sense. What has this environment to offer us, and what have we to do to make it useful? Other cultures were simply obstructions.

The first need was to commoditize the system, to turn it toward the needs of Western commerce. The research approach that grew out of this colonial phase may be characterized in the following ways, which are summarized in Figure 11.1:

1. A preoccupation with the development of commercial (cash) crops for export and an almost total neglect of local food crops unless they offered some prospect for use in the colonial system.
2. A focus on the technical optimization of production within the context of Western science and management.

3. The development of a delivery system to take such knowledge out into the countryside (the extension service). There was no parallel development of any service to investigate broad production and distribution problems in the countryside. The objective was simply one of taking the answers out to the people, not in asking them questions. This was a further manifestation of the supreme self-confidence and arrogance of Western science.

4. The indigenous production system, where it was incorporated at all, was expected to modify itself around the new knowledge regardless of local intrinsic needs and priorities. This was likely to be a serious problem where cash and food crops collided on the same farm. At the same time, there was virtually no serious study of indigenous food production systems or crops since they were, on the one hand, unscientific, and on the other, unprofitable.

5. There was almost no focused social science research into farm production. There were seemingly endless studies of kinship, ritual, and so on. So the organization of production was largely an unknown area.

The Research Station

The logical outcome of the this approach was a phenomenon known as the research station, many of which survive to this day. This was an area carved out of the bush, frequently on the best available land. The research station would develop into an oasis of Western science and technology, capital, and management, where the various factors of production were optimized. At no time did this venture even remotely resemble the hundreds of thousands of peasant farms that surrounded it or the way in which the locals lived. The crops grown were almost inevitably commercial monocultures, and it was quite often the case that these research stations were supported by single enterprise institutions such as tobacco companies, rubber growers, etc. The findings that emerged were then carried forth by armies of extension agents whose job it was to tell — and even force, under weight of law — the farmers what to do. The research station was a microcosm of the dual economy, in which the "oasis" of the research station matched the oasis nature of the settler farms or plantations.

Research in the 1950s and 1960s

As many tropical countries moved toward independence in Africa and Asia after the last war, there began to emerge serious doubts about the future. In general this focused on India, which became independent in 1947. It is hard to imagine now, but a review of, for instance, the Freedom from Hunger campaign literature of the early 1960s points to the coming catastrophe in India, as the country falls further and further behind in its capacity to feed its population. Africa was rarely mentioned in this scenario, and indeed this period was one of unparalleled growth and optimism for that continent. Textbooks on Africa published at that time described the continent as generally "underpopulated". Interestingly, the great famines in India were predicted to occur in the mid-1980s,

Text Box 11.1
Characteristics of Colonial Research

- Preoccupation with commercial commodities at the expense of food crops
- Focus on technical optimization rather than the production system
- Unidirectional approach, generating answers for farmers rather than asking them questions
- "Ethnoscience" not considered worthy of study
- Technical research with no social context

which was in fact the time that India became a food exporter! But there was a general serious concern that the food base had been neglected throughout the colonial period and the new countries were going to face starvation since they were unable to survive by eating the rubber, cotton, tea, or coffee that had occupied most of the best land. In the interim, the United States, under Public Law 480, shipped millions of tons of surplus grain to countries such as Egypt and India, which were slipping further and further behind in their effort to feed their populations.

During the 1960s, the focus of attention shifted toward the food crop base, but the approach remained the same: effective technology transfer. This process, known as the "green revolution", will be reviewed at the end of this chapter. There is no doubt that the green revolution achieved extraordinary increases in the production and productivity of staple cereals. At the same time, the problem of global poverty continued to rise. The number of absolute poor grew, and the industrialization of the developing world proved to be a chimera for all but a few countries on the Pacific rim of Asia. Worst of all, Africa started to fall apart at the seams and turn into the very catastrophe that had been envisaged for India.

Systems

During the 1970s a radical change began to occur in the whole approach to research and development in the tropics. Instead of focusing on commodities, attention changed to the paradox of the burgeoning rural poor in a world of rising GNPs. Indeed it might be fairer to say that concern was growing over the increasing number of the rural destitute. All this was accompanied by soil erosion, deforestation, and desertification of ever-greater proportions. Something clearly was wrong: This was not development.

It was realized that just as the World Bank was publicly stating that it seemed to be possible to have growth without development, so it was possible to increase the gross figures of food and agricultural production without necessarily solving the problems of those at the bottom end of the scale. Success, it seemed, sought out the already successful. Even though technology appeared to come bearing a cornucopia, and even though the statisticians showed global production figures rising ever upward, the picture on the ground was becoming undeniably worse: India's landless population grew by millions, and Africans fell clean through the bottom of the system across the continent. In Latin America peasants became so frustrated and impoverished that the whole area was ablaze with revolutionary war. It was in this context that so many politicians attending conferences cried desperately, "No more research!" They were more than aware that the social consequence of much technical change was chaos and disorder.

What was needed was to integrate the social and cultural, as well as the political, realities into the "neutral and objective" technological approach to research. No more of those oases of technical optimization — now reality must intrude and muddy the waters. The response to this challenge was to move research into areas of interdisciplinarity, and the vehicle for this was the production system. Systems research essentially sought to integrate all those elements that, together, influenced the functioning of the production system. These variables might be as diverse as kinship norms, credit institutions, soil quality, and drought. But it was the essential *interaction* of these variables that gave the system its validity and predictive capacity, as well as helping to demonstrate the potentially destructive nature of culturally dislocative technology transfer. Now, instead of the extension agent and the research station, the system researcher would start by investigating the production unit itself, particularly the modal — or most commonly occurring — unit, i.e., the farm or herding unit. Only then would it be possible to understand and quantify the constraints and possibilities, and more important, appreciate the *needs* rather than the "obstacles" to induced change that were the focus of earlier anthropological studies and the perennial nightmare of the colonial administrator.

All this seems fine and logical. But there were, and are, problems. It is one thing to talk about interdisciplinarity; it is quite something else to operationalize it. People — and scientists are people — are the product of their own environment, paranoias, and prejudices. Most of us (wherever we live) are brought up in an intellectual environment that divides knowledge into artificial but manageable boxes. To cross the boundaries requires an act of faith and humility (and in academia a desire to abandon promotion prospects!), since we are entering unknown territory. For many of us it is easier to remain within the security of our artificial intellectual kingdoms. For that reason, putting a group of people from relevant disciplines together does *not* produce systems research. That requires a catalyst, usually a visionary individual to nurse hurt egos, set aside old prejudices, and demonstrate once more that the whole is more than the sum of its parts.

To help accomplish this, several of the major funders of agricultural research worldwide, such as the Ford Foundation, the Food and Agriculture Organization of the United Nations, Rockefeller Brothers' Fund, and others, established a series of integrated, multidisciplinary centers under a collective umbrella known as the Consultative Group for International Agricultural

Table 11.1. The Centers Operated by the Consultative Group for International Agricultural Research (CGIAR)

International Center for Tropical Agriculture	Cali, Colombia	Rice, beans, cassava, beef
International Potato Centre	Lima, Peru	Potatoes
International Wheat and Maize Improvement Center	Mexico City	Maize, wheat, barley, triticale
International Board for Plant Genetic Resources	Rome, Italy	Genetic research
International Center for Agricultural Research in Dry Areas	Aleppo, Syria	Dryland agriculture
International Crops Research Institute for the Semi-Arid Tropics	Hyderabad, India	Food in the semi-arid
International Food Policy Research Institute	Washington, D.C.	Public policy
International Institute of Tropical Agriculture	Ibadan, Nigeria	Tropical food crops
International Laboratory for Research on Animal Diseases	Nairobi, Kenya	Tick and tsetse control
International Livestock Center for Africa	Addis Ababa, Ethiopia	Livestock production
International Rice Research Institute	Los Baños, Philippines	Rice production
International Service for National Agricultural Research	The Hague, the Netherlands	Research systems
West African Development Association	Monrovia, Liberia	Rice self-sufficiency

Research (CGIAR). These centers provide the intellectual matrix for integration to occur, though some of them are structured along disturbingly conventional lines. Still, an encouraging commitment to a new focus is evident. The centers, and those institutions listed in Table 11.1, combine the social and the natural sciences, have a commitment to the systems approach, and start with a focus on the production unit. Though this is a very late start, and though the crises continue to grow, at least the research basis has now been matched to the nature of the problem. The real challenge now will be to see whether the implementing authorities — the public sector — can demonstrate the same commitment to integration, the producer, and interdisciplinarity. This is examined in Chapter 12. Before then, we consider briefly the way in which one technical "miracle" staved off the starvation predictions of the 1960s, while also demonstrating the potential dangers of a purely "technical" approach. This is something we shall examine again in Chapter 13.

The "Green Revolution"

During the interwar period, remarkable work was done in Europe and North America on crop breeding, which has, for the moment at least, changed the agricultural capacity of many areas. By selective crossbreeding, crops were developed that would mature in a shorter growing season, pushing the frontier of grain cultivation dramatically northward in, for instance, Sweden. In the United States, varieties of cereals, particularly wheat and corn (maize), were bred that were very responsive to fertilizer and managed water regimes. These new crops offered the potential for doubling — or more — the yield of some grains.

Thus, when the Malthusian fears about the future of India started to develop in the 1950s and 1960s, and food crops finally came back on the tropical agenda after decades of official neglect, the existing research on cereal hybridization achieved considerable importance at the global level. It is wise at this point to remember that the research originated in highly capitalized, developed communities since that will be important in understanding what followed.

In 1948 the work conducted in the United States that had helped transform the Dust Bowl areas into the irrigated grainlands that they are today was extended southward to Mexico. By selective crossbreeding of high-yielding grains with local Mexican varieties, researcher Norman Borlaug was able to produce a variety of maize that put most of its energy into producing the seed head and not into growing tall and rangy. To do this the researchers used the "technical optimization" approach mentioned earlier, relating it to the physical environmental conditions to be found in Central America. The results were sensational and started what has been called the "green revolution", for which Dr. Borlaug was quite justifiably awarded the Nobel prize.

The genetic manipulation produces its results only when a controlled environment is produced to nurture the high-yielding varieties (HYVs). Breeding-in certain desired qualities means that these benefits are sometimes achieved at the cost of other genetic qualities that the local varieties have developed over centuries, such as resistance to the local disease spectrum, the capacity to withstand periodic stress points characteristic of the local climate, and adaptations to the microenvironment of soil and plant assemblages. It now becomes necessary to provide the HYVs with a regulated and reliable water supply, otherwise the plant may have much less resilience in the face of water stress than its predecessors, and rather than produce a diminished yield under stress, it may collapse completely. The old varieties might see the grower through good times and bad; the new variety works only in the good times.

Another genetic inheritance that has to be sacrificed to a greater or lesser degree in pursuit of yields is disease resistance. By trial and error and natural selection, the indigenous, traditional gene pool has experienced just about everything the environment has to offer and has bred into itself a broad range of tolerance. The new varieties are, by contrast, completely uniform. If disease does penetrate the defenses, then its effect will be devastating. Further, the new strains have to be thoroughly protected by chemical and entomological means against the threat of disease. Pesticides are an essential part of the HYV scenario.

Text Box 11.2
The "Green Revolution" Package

- Water must be available in the right quantities *throughout* the growing season.
- Pesticides must be reliably available and adequate to meet changing pest resistance patterns.
- Fertilizer must be available at the right time and in the right proportions.
- Hybrid seeds must be available promptly, reasonably priced, and at a convenient distribution point.
- The management know-how must be imparted to the farmer on how to use new inputs.
- Credit must be available to enable farmers at lower levels to participate.
- The market infrastructure must be capable of absorbing the increased flow.
- Technical backup must be in place for irrigation pumps, sprays, tillers and tractors, and other equipment.

Where local gene stocks are replaced by hybrids, it is no longer possible for farmers to produce their own seed. Since seed collected from the hybrids will be on the road to degeneration and will not pass on the qualities of the first crop, it is valueless; the HYVs are not self-replicable. Instead the farmer must buy seed from breeding stations where it is produced by continual crossbreeding, and to do this the farmer must have cash in hand. This is a new form of dependence for many farmers in the tropics, and they must now rely on the efficiency of the market not only to produce the goods at the right time, but to take the farmer's production at a fair price to generate the income necessary to buy the inputs for next year. Unfortunately, the pattern has been for agriculture to provide a bottomless pit, through government price manipulation, to keep the restless urban population happy. Governments have become victims of their own pricing policy and cannot easily break free. The urban population, as in Egypt, expects absurdly cheap bread and turns out in the streets in large numbers should anything threaten this favored position.

In order to keep the flow of water moving around the crop, it is quite likely that the farmer will need pumps, which in turn require fuel. It is certain that there is going to be a substantial capital start-up outlay for pipes, ditches, etc., along with sprays, machinery, stores, and other equipment. It is also essential that the farmer acquire the knowledge to manage the land under the new regime if it is not to be poisoned with chemicals or salinated by overwatering, or if the local water supply is not to be sterilized through eutrophication resulting from nitrogen buildup. It is one thing to introduce this technology in Sweden or the United States, both of which have huge infrastructures of literacy, credit, extension services, technical backup and maintenance, and reliable marketing systems for distribution and collection. Indeed this technology evolved in that context, so it is not unreasonable to assume that it will carry these contextual assumptions with it when it is transferred to some other part of the globe. It is, however, unreasonable to assume that this context exists all over the world.

The Impact of the Green Revolution

After the new varieties of corn came wheat, and then rice. Centers in Mexico and the Philippines pioneered the path, and the impact was nothing short of miraculous. Introduction of new wheat varieties started in India in 1965, interestingly just one year after the United States decided to discontinue the massive shipments of PL480 cereals. By 1982, about 80% of wheat lands in India were under the new HYV strains, up from 50% in 1972 (Figure 11.1). Throughout Asia the new varieties were to be found on 24 million acres by 1970. Within 3 years of HYV introduction, Pakistan freed itself from imported wheat, and Sri Lanka and the Philippines recorded unprecedented harvests. Even during the 1979 drought, India was able to feed its population from its own domestic production — quite a contrast from the catastrophe predicted by the FAO Freedom from Hunger publications of just 15 years earlier. Indeed, in the 20 years following 1950, India *doubled* its production of cereal crops and raised the amount of food available per capita by a quarter. There is no denying the technical achievement of this revolution, and like most revolutions, it has ushered in a different world. The question is, can it last or did it just buy breathing space?

The Social Context

It is clear that in order to benefit from the green revolution the farmer has to acquire it as a "package". It does not work in increments. Growing the HYVs without pesticides means risking the loss of the entire crop to insects and viruses. Growing it without supplementary irrigation may result in the whole crop withering rapidly under stress. The farmer is therefore faced with the task of capitalizing the revolution, and in India this represents about 7 years' total income. This would be reasonable if there was a credit system in operation that served small farmers at realistic rates. Unfortunately, most banks look upon small farmers as credit risks. In fact the reverse is true; it is the middle and larger farmers who usually default. In Zambia, for instance, a small-farmer credit program was established to meet the needs of this group. However, out of 25,000 loans, only six were to small farmers. The reason given by the credit authority was that the small farmers were illiterate and had a hard time filling in forms, they had no postal address, and they had no security to offer the bank against the loan (because they were poor!). Furthermore, the overhead cost of so many tiny loans was, said the bank, prohibitive. So, the small farmer had to be well off to get the loans, and that is exactly what happened — only the richer farmers benefitted. However, the better-off group repaid less than 50% of what they borrowed. In Malawi and Bangladesh, two of the few countries that do make small loans to peasant farmers, the repayment rate is in the high 90% region, rendering the arguments of the Zambian bankers totally spurious. The bankers were looking for an easy life and living out age-old myths about the hopeless, irrational peasant.

Small farmers first have to cross the credit hurdle. If there is not a functioning formal system, they must throw themselves on the informal system with its punitive interest rates and lifelong indebtedness. Next, farmers become reliant on the vicissitudes of the market system to sell and buy, for this is a commercial venture. There is no way that the green revolution can enter a subsistence system; it simply *has* to be commercial. Now the specters of venal middlemen and inept parastatals loom on the horizon.

What has happened widely in Asia and Latin America is that the green revolution has been adopted by those who have access to the capital, namely the better-off farmers. As they introduce the package, their yields increase accordingly, while yields on the smaller, traditional farms remain the same. There is an increase in the flow of production into the market, which tends to depress the price of the wheat or maize. That is fine for the urban dweller always on the lookout for cheap food. However, it *reduces* the income of the small farmer, and all too often the farmer is forced to sell the farm to one of the better-off green revolutionists. So the net effect of this process can be to polarize divisions in society. Of course this is precisely what happened in much of Western society with, for instance, the enclosure movement, and the various scale economies of farming in the last 150 years. There, however, alternative employment was offered in the burgeoning urban/industrial context. In much of the developing world, the job opportunities are simply not there to absorb the refugees from the green revolution. In India the number of landless peasants has grown to over 30 million, and the green revolution is of little benefit to people who cannot buy the cereals and who have lost their own means of production.

The package outlined earlier, which provides the essential "life support" for the green revolution, is heavily dependent on hydrocarbons for the production of fertilizer, pesticides, plastic pipes, and fuel for pumps and other machinery. The great phase of expansion of the green revolution was between 1965 and 1975, and at that latter date the OPEC nations started a process that was to raise hydrocarbon energy costs by over 1200% by the end of that decade. This changed the equation, seriously pushing up the costs of inputs. At the same time, however, the increasing production of cereals threatened to diminish the unit value of the output. There has been far less use of fertilizer as a consequence, so that now less than one third of the required amount of fertilizer to allow the HYVs to produce their optimal yields is used. In the rich countries, the green revolution seems to have run most of its course, leaving little marginal growth in the system as a result of greater fertilizer use. It is always possible that genetic engineering will alter the potential of the gene stock available to us, but right now the developed nations seem to be atop an "S" curve as far as the green revolution, as we know it, is concerned.

There are several areas of concern regarding the future of the green revolution in the developing, tropical world:

1. We have seen the extraordinary rate at which the HYVs were disseminated across Asia and Latin America. As they spread they displaced the indigenous crop varieties that had formed a gene pool uniquely adapted to the local environment and left instead this vulnerable, capital-dependent monoculture of hybrids. There is real concern now about the loss of that gene pool and the reduced options its loss or reduction offers agriculture in the future. Elsewhere in the book we lamented the loss of traditional "ethnoscience". The same is true for local crop and plant assemblages. Some efforts are now being made to form gene banks of traditional varieties.

2. To meet the recommended optimal use, farmers in the tropics should be using three to four times as much nitrogen supplement as at present. If India alone were to do this, it would consume about 30% of current global production. There is also a question of how much additional nitrogen the world's cultivated ecosystems can take, since so much of it ends up elsewhere in ponds, lakes, rivers, etc.

3. The package has made many farmers dependent on networks over which they have no control. The health of the green revolution will depend on the efficiency of these market, credit, advice, and other networks.

4. The green revolution almost completely bypassed Africa. That continent has a minimal amount of irrigable land and its staple crops, especially the millets, did not receive the attention given to rice, wheat, and corn — all of which are valuable temperate crops as well. Over the last two decades, the average level of food production per head has actually been *declining* through most of Africa. Only now has research begun into the African staples, but already Africa has been dramatically transformed from a food exporter 20 years ago to a food importer whose import demands double every decade.

Where Next?

There is still considerable potential for greater yields in the tropics from the use of more fertilizer in conjunction with HYVs and that will buy a little more time. However, there is real concern about a system that relies so much on chemical additives, imported hydrocarbons, and the world market system. Is this in the realm of sustainability? There is growing interest in researching the following ecodevelopment principles:

1. Mixed tree/crop plantings. Trees can bring nutrients and moisture from deeper horizons than the average food crop. Trees and bushes also reduce wind velocity and the desiccating effects of unimpeded wind passing over bare soil surfaces. Trees, furthermore, provide fuel on a renewable basis.

2. Genetic engineering can work in a low-capital, as well as a high-capital, green revolution environment. More rapid responses to sunlight, tolerance to moisture stress, and selective disease resistance are all qualities that can be bred in, as well as bred out. This frees the farmers from pesticides and fertilizers that, for the most part, they cannot afford anyway.

3. Organic farming has the capacity to produce impressive increases in yields, but has been regarded as peripheral and eccentric by mainstream researchers for many years. On the other hand, well-managed organic farming produces yields that compare favorably with green revolution figures.
4. The green revolution was born in the West, and in that part of the world there is real concern about the buildup of chemicals in the environment. Americans, for instance, use more fertilizers on lawns and golf courses than are used throughout the whole of Africa. It may well be that this increased nervousness and consciousness may spur a new approach to researching sustainable agriculture. That in turn may prove to be the second wave of change.
5. The green revolution bought the world some time to think about its future and research its options. It did not do this everywhere, as is clearly demonstrated by the parlous state of Africa that it bypassed. Nevertheless, it will not solve the problem as long as the old Malthusian dilemma continues and population swells to gobble up the benefits of increased productivity. To match growing demand by using only the green revolution approach, we could end up poisoning the planet.

USEFUL READING

- Allen, John, Ed. *Environment 90/91* (Guilford, CT: Dushkin Publishing Group, 1990).
- Consultative Group on International Agricultural Research. "*1986 Annual Report.*" Washington, D.C. (1987).
- Glaeser, B., Ed. *The Green Revolution Revisited* (London: Allen and Unwin, 1987).
- Hossain, M. "Nature and Impact of the Green Revolution in Bangladesh," International Food Policy Research Institute, Social Report 67, Washington D.C. (1988).
- Plunknett, D. L. and N. J. H. Smith. "Agricultural Research and Third World Food Production," *Science* 217:215-220 (1982).
- Wolf, E. "Beyond the Green Revolution: New Approaches for Third World Agriculture." Worldwatch Paper #73, Washington, D.C.: Worldwatch Institute, 1986.

CHAPTER 12

The Administrative Trap

In the same way as knowledge was shown, in Chapter 11, to be traditionally divided into disciplines, so public administration has been conventionally divided into sectors. These are usually functional units, such as ministries or departments, which take care of discrete aspects of managing the public domain. These separate units of the public sector may well define the "problem" quite differently according to their segmented, sectoral view of the world. Where research information on a problem is handed down in a highly integrated way, it presents a problem for the compartmentalized bureaucracy that then has to disaggregate it before it can make "use" of the information. In these circumstances it becomes extremely difficult to mount an effective public response to large-scale and highly integrated ecological situations even if the political will appears to be present. This administrative dimension can become a trap that distorts and limits action, and it is frequently overlooked in the framework of analysis with which we approach environmental degradation and loss.

This administrative dimension is particularly important in the tropics because so many people depend directly upon the natural resource base for their livelihood. In most of the countries that lie in the tropics, the governments are much more *interventionist* than they are in the developed, temperate lands. The administrative aspect of environmental management is also important because it is the area of public life where policies and ideas are implemented. The structure of the administration may well form an effective block to successful implementation of sound management principles and conservation.

It is useful to start by looking at a simplified model of the typical sectoral structure of public administration in most tropical countries (Figure 12.1). The names of the units may vary somewhat, but the overall structure is fairly

---

Text Box 12.1

Characteristics of an Ecosystem	Characteristics of Administration
• Common denominator = energy • Total integration of all elements • Bounded by physical laws • Tends toward stability • People part of the system	• Common denominator = development model • Disaggregation • No absolutes, human values • Tends toward change • People are the system

---

Text Box 12.2

Factors Emphasizing the Need for Administrative Reform in Tropical LDCs

- The poorer countries in the tropics are undergoing extremely rapid change, particularly in terms of their population growth rates, and great destruction may be brought about in a very short time.
- The environmental problems are vast and involve a wide array of physical, social, and economic dimensions demanding action on a grand scale.
- There is a need to think beyond the short-term horizon of political and economic expediency.
- There is an urgent need to make the best use of scarce human and capital resources.
- Research flexibility and policy flexibility must be matched by administrative flexibility.
- We must find an alternative to economic and political crisis management in order to manage the environment.

---

representative. The figure also shows the points at which innovation may take place, and these are discussed later in this chapter.

At the top of the hierarchical structure are the policy organs, sometimes legitimized by popular participation, sometimes not. Here, in theory, the will of the people is expressed, their hopes and ambitions are enshrined in policy and legislation, and the goals and principles are established for the future. The policy and legislation will guide the many actions that the public sector will carry out and will provide encouragement and sanctions. Of course, if the political will to enact sound environmental management is not there, we may not expect to find it expressed further down the system. However, the situation is not quite that simple because many of the governments in the tropics face extreme pressures to provide short-term solutions to extreme economic hardships. Long-term sustainability of the environment may be both logical and desirable, but it may not be politically feasible in countries facing enormous debt burdens, rural-urban migration, unemployment, and political instability. The horizon of most politicians is determined by the next election, pressures from constituents for answers *now*, and a need to deal with situations on a

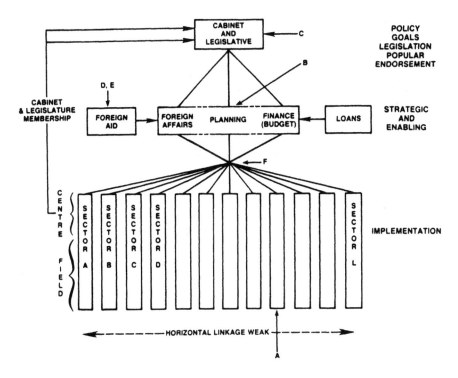

Figure 12.1.   A typical sectoral model of administration. Key to innovations for envi-
ronmental action: A, ministry of the environment; B, plans and programs;
C, policy, legislation, and lobbying; D, foreign aid projects; E, UN action;
F, regions and problem areas.

crisis-management basis. It is easy to blame politicians in developing countries
for having insufficient environmental awareness, but sometimes we forget to
ask, "What political price has to be paid for being a conservationist in the
economic realities of these countries?"

Below the policy level comes the strategic and enabling level consisting of
three main parts. The first of these is planning, since most countries in the
tropics have had a planned economy in which the public sector lays out its
programs, usually on a 5-year basis. This is indicative of the interventionism
and control mentioned earlier. However, the plan will be only as good as the
policy that directs, guides, and monitors it. Unfortunately, most plans have
adopted the same disaggregated, sectoral approach as that on which the government is structured. This is not surprising because the plan outlines action for
the different existing arms of the administration. Also at this level is the finance
authority with the responsibility for drawing up the budget. Again there is a
problem, because most budgets relate to the structure of the public sector, not
to the nature of the major problems it faces. Most of the allocated funding
comes in response to requests from the individual, disaggregated sectors.
Furthermore, most budgets are constructed on an annual basis, obstructing the

---

Text Box 12.3
The Administrative Trap

The "trap" results from the fact that governments and bureaucracies may be organized into departments, budget lines, policy-making machines, etc., which are not determined by the *nature of the problem*, but as a result of some other intellectual, political or other process. Thus, when the public policy and administration machine confronts large-scale, highly integrated environmental problems, it sees these problems through distorting filters that can and often do:

1. Redefine and distort the nature of problems, since these have to be defined by someone, or some department, presumably, with resources and responsibility to do something about the problem. The nature of the "question" may be determined by the nature of the answer, rather than vice versa. Government departments ask questions framed by their "terms of reference".
2. Similarly distort the nature of the solutions, since the various agencies will tend to treat symptoms of their arbitrarily defined problem, rather than the central issue.

---

ability to take a longer-term perspective on sustainability of the environment. The third institution at this level is the ministry or department of foreign affairs, which may seem to be a curious player in this ensemble. It is present because so many of the governments rely on foreign aid and loans for major capital projects, and these are part of a *political* relationship that is a dimension of foreign relations. At this point it is worth mentioning that having so many influential players outside the national picture provides many opportunities for pressures to be brought to bear on policy and programming, which are not present to the same degree in the developed, temperate countries.

The third level is that of implementation. Action is usually reserved for the various organs of the line ministries, or departments. They act in accordance with instructions and funding from above, though they may have field representation. The various sectors are characterized by this vertical pattern of communication and not by any form of horizontal linkages. Integration is, therefore, extremely difficult. Where it exists, integration has to occur at the secondary or primary levels, but this rarely occurs.

Typically, each ministry makes its own proposals for the plan and requests to the finance ministry in accordance with the ministry's own rather narrow, and often technical, view of the world. It is clear that this structure is going to present many difficulties when it comes to tackling anything as highly integrated as ecology.

The Administrative Trap in Action

Figure 12.2 shows how this structure works, or rather doesn't work, when faced by a major environmental problem. The case used here is the

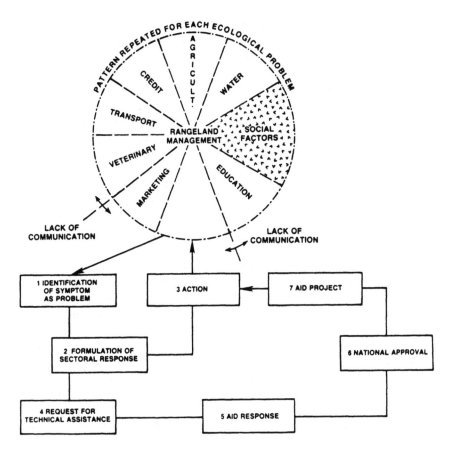

Figure 12.2. The administrative trap — a sectoral spproach to ecological problems. Key: ⬚⬚ largely neglected; – • – boundary of ecological problem; – – sectoral boundary (defined by administrative divisions).

management of unimproved, communal rangelands, though many others might have been used such as desertification, the destruction of the rain forest, etc. The boundary of the circle represents the totality of the environmental problem, all the elements that together and interactively constitute the issue to be tackled. Many of the pastoral areas in the world are being severely degraded, causing hardship for the people who live there and a loss of resources from the public trust. In general, improvements in groundwater technology and veterinary services, enabling more animals to survive, have led to rapid increases in stock numbers. Since these animals often form part of a subsistence, insurance, and social prestige system, there are strong pressures to retain the additional animals on the rangelands rather than to sell them into the market.

Within the circle it is possible to subdivide the elements of the problem into areas of conventional government authority: water management, stock marketing, communications, veterinary services, employment, and so forth. No government department alone has the authority or competence to deal with all the elements that, together, cause the rangeland deterioration problem. It is particularly hard for departments to come together and reproduce the problem administratively because of the traditional separation of responsibility and budget discussed here.

What happens then is that each branch of the administration sees only those elements of the problem that fall within its terms of reference. In other words, it focuses on the symptoms of the malaise and not on the sickness itself. Indeed, many of the component ministries may not view the phenomena they observe as part of a rangeland management problem, but as an "animal health" or "water provision" problem. It might be argued that if all the arms of the administration do their jobs effectively, then the problem will have been tackled in its totality. This is not so, because action in one area without reference to the consequences in other sectors is as likely to move the pressure around, or accentuate it, as it is to cure it. Ecology exemplifies the old adage that "the whole is more than the sum of its parts."

Having identified the symptom as the problem, the ministry will now proceed to request the means to tackle the symptom. A good example of this has been the provision of water points for grazing communities in many parts of the world. The indiscriminate sinking of wells will allow many more animals to survive droughts and other hardships pressing on the rangelands. This is particularly true, as we have observed, for people who depend on their animals for subsistence. These wells may open some areas of grazing that were closed because of an absence of surface water, but the provision of water with no thought for what will happen to the grazing resources leads to serious erosion problems. These instances of erosion may be seen in many parts of the world. One particularly bad example is to the southeast of the Aral Sea in former Soviet Central Asia. Here, on satellite images, one can see large circles of degraded land around the water points, like some terrible disease spreading over the face of the landscape. The initial problem of water shortage has been *shifted* to the much more imponderable problem of erosion.

One extension of this administrative trap is that of foreign technical and capital assistance. It might be thought that the growing environmental consciousness in the West would lead to a shaping of technical advice and loans into a more sustainable package. The difficulty is that most of the projects for which assistance is requested are identified by the recipient governments, simply providing for an extension of the administrative trap. As we shall observe later, many arms of technical assistance have become aware of the damage they are causing through financing and have made policy moves to prevent this. Unfortunately there is a tension between the sovereignty of the recipient country and the possibility of interference by outsiders. Consultants are trapped by their terms of reference, which may well be drawn up by their employer governments.

It is important to make one more observation. Within the circle in Figure 12.2 there is an area of social and cultural factors, which are just as important as the more physical and technical ones. In the past these have either been neglected or treated superficially, resulting in the technical and "economic" dimensions being enacted within a contextual vacuum. This explains the failure of more livestock development schemes than anything to do with technical or economic failures at the appraisal or implementation level. Of course it is easy to blame failure on the people "who simply do not understand what is good for them." The reality is often that the technical specialists have little idea of what the people think or want.

Initiatives to Change the Situation

Ministries of the Environment

To those who subscribe to the administrative trap argument, the creation of a "super ministry" of the environment might have seemed like a logical administrative response. With one step it seems to create an arm of government that is defined in the same way as the problem. It is clearly a recognition that the environment is "on the agenda" and that the government is apparently trying to do something about it. Since the Stockholm conference in 1972, there has been an explosion of the number of these departments and ministries. In 1972 there were 10, by 1974 there were 60, and today there are more than 110. There are problems, however, which may be outlined as follows:

1. The ministry is allowed to function effectively only insofar as its recommendations are not at variance with the political expediencies of economic growth. At the moment this is seldom the case, and so the longer-term considerations are likely to be ignored or compromised.
2. A ministry of the environment is like any other ministry unless it has some powerful all-embracing legislation that will force other ministries to pay attention to what it says. In view of the wide brief that any environmental organ must have, it is quite likely that it will be seen as a threat to the power base of the existing structure. Established ministries may take steps, in this case, to neutralize it by ensuring it does not have cabinet status or that there are other ways to bypass its rulings. It is going to be hard to fit an institution so dependent on horizontal authority into a vertically structured system.
3. Much of the existing "environmental" legislation already belongs to other ministries, and loss of this legislation to another ministry diminishes the authority of the first ministry.
4. Since environmental problems embrace, by definition, almost every aspect of life, is this ministry going to need expertise in *every* area of the economy? If it does, then does this mean an enormous duplication of personnel? For instance, if environmental impact statements are to be prepared by anyone other than the line ministries, can they be trusted to police themselves?
5. What, precisely, is the role of this ministry? Is it to be a policeman, counselor, inspector, policy formulator, or regulator?
6. Is this unit going to use up valuable financial resources needed for what is perceived as the "development effort"?
7. What is the relationship between the ministry and the planning process? Does it take a lead or does it do follow-up work as a sort of inspector?

Most critical is the perception of the ministry in the context of the "development effort". The United Nations Environment Program (UNEP) noted the following in 1982:

> Many countries have created environmental secretariats, ranging in scope from the national level down to the sectoral planning sections. However, it is not clear how effective these are in influencing the much stronger, and longer-established economic planning sections.[1]

In the absence of any clearly stated national policy and legislation regarding the relationship between environment and development, it is difficult to see how these units will function at all. Without a clear policy and legislation, they may be created

> in order to satisfy the minimum requirements of bilateral and multilateral donors [so that] they give only pro forma attention to environmental impact assessment. Environmental units in government [are] often only window dressing . . . placed at low bureaucratic levels, poorly funded and understaffed. The powerful economic planning units ignore environmental concerns and [regard] the internalizing of damage costs and degradation as difficult, time consuming and not useful.[2]

It appears that the only way a ministry of the environment is likely to survive in the segmented world of administration is if it is placed under the executive wing and if it is made undeniably clear that the political will is there to see that it works. At this level it can operate to ensure that environmental principles get into the development process from the beginning, not as an afterthought.

Problem-Oriented Authorities

When the United States faced the enormous regional problem of the flooding and poverty of the Tennessee Valley catchment, it reacted by creating a special administrative unit, the Tennessee Valley Authority (TVA), which encompassed existing state and local boundaries. Within the TVA special laws and economic packages operated to bring the river under control and improve the livelihood of the people living there. Although it performed a remarkable job, it has often been criticized for being "outside the system", "privileged", and so forth. India created the drought-prone areas administration to deal with regions of almost legendary hardship. Such innovative responses might be considered elsewhere; for instance, in the tropical rain forests. The problem would be the creation of parallel governments, and the resentment that this engenders within the established system. Furthermore, the affected areas are often remote, and people posted to them feel they will be overlooked in the promotion and power game that requires presence at the center. Since these administrations are set up to deal with specific, large-scale crises, they should have a fixed life span coupled with performance in dealing with the issue at hand. This makes them far less threatening.

Much the same argument may be leveled at regional administration and decentralization efforts. In theory this devolution gives every part of the territory an "equal chance" in the process of appraisal. In reality what often happens is that the regional authority merely reproduces the structure of the national authority and there is no additional integration. Furthermore, many of these instances of decentralization do not involve a similar distribution of budgetary power, allowing the regional initiatives to be overruled. Lastly, it is often difficult for poor regions undergoing environmental degradation to find the skilled personnel needed to effect administration at these levels. What may happen is that the regional authority simply adds another level of bureaucratic obstruction to the process.

Lobbies

Much of the environmental change at the administrative and legislative level in the West has been the result of powerful lobbying by interest groups. This has reached the level of the formation of the "green parties" in Europe, which have directly entered the political process. In the tropics there are problems with this type of change. Generally there is not the well-developed middle class that has spearheaded the movement in the West. These often well-educated and economically secure people have the resources, political contacts, and "clout" to influence change. The people most affected by environmental degradation in the tropics are generally the poor, and they possess the smallest voice in the process of change. They also lack the resources to mobilize. If we look at the tropical rain forest in Brazil, we find a situation where those organizing to protest the rape of this resource are subject not just to indifference, but to violence, intimidation, and murder. It is not surprising that so much of the outcry about the tropical rain forests comes from concerned groups in the West. They may have good reason to be concerned because of global warming. On the other hand they may help publicize the case for the powerless. The danger is, and this is evident in Brazil, that they are seen as interfering and impinging on the sensitive area of sovereignty. The urban middle class and the landholding class are deeply entrenched in the benefits they receive from the status quo and are a powerful instrument resisting change. The force for real change will most probably have to come from inside rather than outside the countries, and we are beginning to see the mobilization of groups such as the *Chipka* tree protection group in India and the rubber tappers of the Amazon. Outsiders should focus their attention on assisting these groups according to each group's manifesto, rather than trying to write the manifesto for them or force the pace and endanger the movements.

Projects and Programs

One characteristic of much of the "development" effected by the public sector in tropical less-developed countries (LDCs) is the use of projects. These are not normally a component of public action and expenditure in the developed world. A project is a discrete investment package with goals, a fixed life, a geographical location, and a separate budget. The reason that public activity takes this form has much to do with the external funding of so much capital expenditure in developing countries. A convention has arisen that funds are normally allocated not for general revenue support or for recurrent costs, but for identifiable capital projects (dams, reforestation schemes, feeder roads, etc.). It might appear that this model offers an excellent opportunity for integration and the inclusion of sound environmental management principles. The problem, however, is that many of the projects are identified by and commissioned for the sectoral arms of the recipient government. The administrative disaggregation is done before the project is born. Since many of these projects are externally funded, the opportunity exists for donors and lenders to require environmental considerations as part of the appraisal process before lending the money. As we shall see elsewhere, there is a move on the part of the U.S. Agency for International Development and the World Bank, among others, to include this requirement. It is a sensitive issue since it may raise the cost of a project or slow down its design and clearance, and it may look like outside interference in the affairs of others. On the other hand, a donor or lender has both the right and responsibility to see that loans and grants are not used in violation of principles that are adopted by the donor/bank or are the legislated norms of the assisting country. Donor agencies should never be in the position of telling the recipient country what to do. But they may perfectly well tell the country what they will not fund.

Above the level of the project should be the program level. This is a broad statement of an area of policy priority expressed by the government. It is more than a project since it will not necessarily include concrete items, and it is not necessarily restricted to one place. It is a broad avenue that will generate the projects. All too often this level is missing, leaving nothing between broad and nebulous policy statements and the considerable detail of the national development plan. It is at this program level, and of course in policy, that the expression of environmental management principles would be best expressed.

Conclusions and Guidelines

1. It is important that in the overall environmental effort we include the management structure and administrative system as a variable. Too often this is taken as a "given", and sometimes it may well be an essential part of the problem. Few studies of environmental change include this dimension implicitly.

2. As we discussed in the chapter on development, a climate must be created that enables governments to act in the long term. The present economic situation does not encourage this; indeed it prevents it.

3. We have to recognize that many of the environmental problems are caused by or exacerbated by poverty and polarization of wealth. While forces outside the national domain can do little to change this, they should not, through assistance, support and sustain these unequal systems.

4. We must evolve — and both the OECD and the World Bank are working on this — a workable methodology for internalizing environmental considerations into both national policy and planning. It is not enough to tack on an environmental impact analysis as an afterthought. A real environmental policy would generate projects, not just react to them. As the OECD noted:

> Environment is now beginning to be seen not just as an additional subject, the examination of which has to be added woodenly on to traditional development considerations, rather it is increasingly seen as a whole new approach to development and gives greater weight to the sustainability of results and to the costs of the destructive side effects.[3]

5. We need to evolve a more effective forum for bringing together the technical, political, and administrative dimensions of the problem. At present the UNEP affords the possibility of this and has done remarkable work in the Regional Seas Program in the Mediterranean, for instance. However, there is still an enormous dysfunction between the pressures faced in everyday life by politicians and the characterization of the politicians as being indifferent to environmental considerations. There is a difference between not caring and not being able to do anything realistic to change things.

REFERENCES

1. United Nations Environment Program. *Review of the Major Achievements in the Implementation of the Action Plan for the Human Environment* (Nairobi, Kenya: United Nations, 1982).
2. Carpenter, R. A. and J. A. Dixon. "Ecology Meets Economics: A Guide to Sustainable Development," *Environment* 27(5)6-32 (1985).
3. Organization for Economic Cooperation and Development. *Aid and Environmental Protection Ten Years After Stockholm* (Paris: OECD Development Committee, 1982).

OTHER USEFUL READING

• Baker, R. "The Administrative Trap," *Ecologist* 5(7): 247-251 (1976).
• Baker, R. "Institutional Innovation, Development and Environmental Management: An Administrative Trap Revisited," *Public Adm. Dev.* 9(1):29-48, 9(2):159-168 (1990).

CHAPTER 13

Environmental Degradation in Kenya: Two Conflicting "Explanations"

This chapter examines in detail the nature of a serious paradox raised earlier: Why is it that, despite a rapid growth in research, institution building, training, and investment, we are witnessing an acceleration of environmental degradation through salinization, soil erosion, catchment destruction, and so forth? One conventional approach is to treat the environmental issue as the problem and seek a technical solution. The repeated failure of these technical solutions is then usually attributed to some form of aberrant behavior such as irrationality, economic perversity, ignorance, or tradition. If we step back one pace and pull the policy-and-decision-making system itself into the analysis, then the environmental problem is revealed to be only a symptom of broader issues: the political economy and access to resources. The technocratic approach to land degradation is conveniently used by those wishing to avoid political realities, and in truth, it will never solve the problem of environmental destruction in places where the environmental damage is a secondary result of a broader sociopolitical crisis. However, the technocratic approach remains popular simply because it does allow those people at the top to avoid reality, shift blame, and appear to be doing something. In reality it usually allows the situation to get worse until political conflict offers the *only* means of resolution.

Although this critical and sometimes deliberate misperception of the problem operates principally at the national level, it tends to be reinforced at two other levels. On the one hand, the global conferences, such as those organized by the United Nations on the environment, agriculture, water, habitat, etc., stress the technical or "mismanagement" aspect of the situation because to do otherwise would be seen by many member countries as an infringement upon their sovereignty and internal policy. Furthermore, the position papers presented at these gatherings are usually prepared *by governments*, so that the government system is subsumed and is not made an object of study. It would be a rare government indeed that would stand up at an international forum and identify itself and its policies as the cause of environmental breakdown! So the direction and organization of the government system is usually excluded from the frame of analysis.

The second agent of reinforcement has tended to be aid and technical assistance. It is essential to remember that most forms of technical assistance are offered in response to a request from the recipient country. It is up to the country to identify and define the problem, and once more it is unlikely that the requesting government will identify itself as the problem. Aid agencies are also very sensitive to sovereignty issues, and bilateral aid (the principal component) is most often part of a political relationship, making such sensitivities very central. So, once more, the political and social status quo become implicit in the relationship. The very term "technical assistance" implies the idea that the cooperation is within a framework of neutrality or objectivity. There is rarely any awareness of the historical process that has produced the conditions leading to environmental destruction. There will therefore be a tendency to treat the physical symptoms of the problem rather than address the wider causes.

Although Kenya is being used as the case study in this chapter, the comparison between the two approaches is one that will be recognizable to researchers in many countries (those in Central America, for instance). The general nature of this conflict of approaches raises serious questions for those in the environmental sciences. Frequently, researchers see the political dimensions of the problem on which they are working as something totally divorced from their technical expertise (a soil is a soil is a soil . . .), as well as being something rather disreputable that interferes with the scientific objectivity of their efforts. It is important that all such people realize that their work cannot be separated from the political and social context in which they work. Otherwise they may very well end up providing palliatives that will actually make the situation worse both physically and politically. The idea of "I am only doing a job" or "the client has to deal with those other issues" cannot be reconciled with any form of intellectual honesty. Worst of all is to say, "I cannot involve myself in the political circumstances of other countries." *All* technical assistance operates within a political paradigm, and it would be strange if it did not. The technical consultant becomes part of the problem just by being there. Ignoring the sociopolitical context is an inherently conservative approach stressing the status quo simply by excluding the political and administrative system as a causal element.

The Environment and the Loss of Resources in Kenya

Kenya is divided into two broad, contrasting, ecological zones (Figure 13.1). Four fifths of the country is classified as marginal, semiarid, or arid,

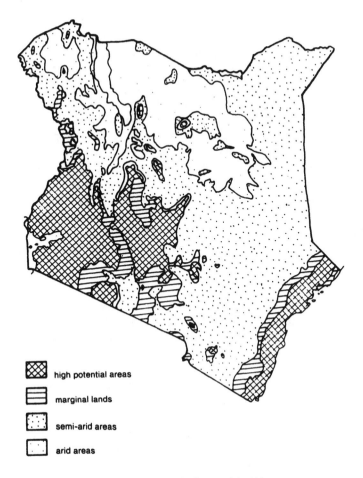

Figure 13.1. Ecological potential of Kenya.

accommodating one fifth of the country's inhabitants. The remaining four fifths of the total population of 24.3 million are compressed into the high-potential agricultural areas. There is a tendency to extend this physical division into a typology of environmental degradation so that losses in productivity and the natural resource base in zones 2, 3, and 4 are attributed to "desertification," while in the remaining areas they are attributed to soil erosion and deforestation. The difference, however, is only in the end state of the process. The causal mechanism may well be the same. In fact this division serves only to confuse the issue, because it somehow suggests that these are different "problems".

The evidence of environmental degradation is widespread and often alarming in Kenya and is often classified into the following categories, although at the physical level they are intimately linked (for instance, forest catchment destruction may lead to soil erosion).

Soil Erosion

Soil erosion is undeniably evident on slopes exceeding 15%, near river banks (despite laws preventing cultivation in these areas) and in the marginal land (zone 2), where cropping practices, usually involving maize, have been initiated by migrant peasants. Most of this soil erosion is water induced, but in the semiarid regions, wind may play a significant part once protective vegetation has been removed by overgrazing. The general effects of erosion have their parallel in the declining yields of the basic food crops since 1970. Denuded soils threaten the Tana River hydro scheme, where accumulations of silt are building up behind the dams and rapidly wearing away the blades of the electric turbines, reducing the expected life of these costly investments. An estimated 3 million tons of silt per annum is collecting behind the Gitaru Dam, and turbine blades have had to be replaced well ahead of schedule. Some rivers, such as the Perkera, have ceased to be perennial, their regimes having been totally disrupted by surges of water from bare, deforested land, and the huge buildup of silt along the bed.

Deforestation

The gazetted, or officially designated, forest reserves of Kenya were estimated by the Forest Department to cover 3.5% of the country in the early 1980s. Much of this forested land serves a vital function as a water catchment for Kenya's major rivers. However, since 1970 some 15,000 acres have been legally excised from the reserve. There is no figure for the illegal encroachments, but an analysis of Landsat imagery in 1980 showed that the actual forest area was 2.5%, not 3.5%. Much of the loss is attributed to uncontrolled clearing for small farms.

The insatiable demand for fuelwood exerts pressures on the forest. It is estimated that Kenya's annual fuelwood requirement is 30 million cubic meters, far in excess of the legally allowed supply of 200,000 m³. The UNESCO Integrated Project on Arid Lands, situated in northern Kenya, noted in that area ". . . a decline in indigenous wood cover, a lowering of the water table, and the spread of sand. The clearing of the forests on the mountains of northern Kenya has destroyed the river regimes and threatened the livelihood of the people."

Overgrazing

The impact of overgrazing in the drier areas is so advanced that the affected areas show up clearly on satellite imagery. It is most concentrated around settlements and water holes where animals must congregate regularly. This downgrading of the rough pasture makes a risky lifestyle even more perilous, and in 1975, 25% of the livestock in the Eastern Province died. Overgrazing of the semiarid lands leads to desertification, whereby marginally usable land become totally useless.

The Technocratic Approach to the Problem

A summary of the main characteristics of the technocratic approach appears in Text Box 13.1.

Text Box 13.1
The Technocratic Approach

- Ahistorical, recognizing no historical antecedents
- Deals only with symptoms of much worse, fundamental, causal processes
- Can lead to increasing polarization in the economy and society, penalizing people for actions resulting from their own poverty and hopelessness and eventually precipitating political crisis
- Maintains a facade of technical objectivity and an appearance of concern
- Those perpetuating the approach at the policy level have a vested interest in its use, though it is quite possible that they genuinely believe that they are acting in the "best interests" of the country as a result of the model of "development" to which they adhere
- In the context of political economy, this model almost never proposes change in anything other than "land management" terms

---

PROBLEM	SYMPTOMS	CAUSES	SOLUTIONS	CONSEQUENCES
Kenya has an environmental crisis	Desertification Deforestation	Overpopulation Overgrazing Overcultivation	Family planning	Lack of response
	Soil erosion	Ignorance Tradition	Education Change attitudes	Inappropriate knowledge = frustration
	Catchment loss	Culture Inapp. practices	Demonstrate new ideas	Short-term palliatives
	Silting Decline of rivers Decline of food production	Lack of environmental awareness Inadequate legislation Institution weaknesses	Environmental education and E.I.A. Tougher legislation Integration Ministry of the environment	Rationalizing oppression Oppression and polarization Protect environment against people New and more efficient ways of avoiding problem

Figure 13.2. The technocratic perception: "environmental protection".

In the technocratic approach, the *environmental crisis* is the baseline and everything derives from this viewpoint; that is, there is a crisis with the environment, not with any other aspect of the system. It is the environment that is sick and must be cured. Once the problem has been defined this way, the whole sequence of events shown in Figure 13.2 will most likely follow.

Physical parameters such as desertification, soil erosion, etc. are seen as the symptoms or indicators of the problem. Success or failure in dealing with the problem will be measured in terms of how far the authorities are able to reduce, halt, or reverse the physical damage. This involves the extension of more effective control by the central authorities over the natural resource base. Reducing the pressure on the environment may, however, involve increasing the pressure on the people causing the damage. This seems logical in a context where people are willfully damaging the environment, but as we shall see, pressuring the people makes sense only where they have an *alternative*. Those in authority may justify their actions by pointing out that they act as trustees maintaining the viability of the natural environment for future generations.

When it comes to "causes", in Kenya — as in Guatemala or Nepal — the issue of population is identified. This is manifested throughout the press and in government statements by expressions such as:

> Kenya will enter the 21st century with about 34 million people. . . . Most of the problems that will continue to face this country . . . are closely related to the present high population growth rate of about 4% per year — a situation that is referred to in some circles as the "Rabbit Syndrome". . . . Such a population growth rate, doubling after only 17 years, would continue to complicate the planning efforts, quicken the depletion of scarce resources and undermine economic prospects. It is undesirable.[1]

As early as 1977, the Ministry of Lands and Settlements stated that by the year 2000 the "surplus" rural population would be 6 million. In the marginal lands the growth estimate, because of migration, is estimated at *10* times the national average. Clearly, the government believes that it is confronted with an "overpopulation" problem. The scale of the population problem is illustrated in Figure 13.3.

The second major "cause" connected with this scenario is that of "ignorance, tradition, and attitudes". These are connected with the "Rabbit Syndrome" because overpopulation is seen as the result of mindless procreation, giving no thought for the future viability of the natural resource base. Similarly, the destruction of much of the marginal land is seen as a result of "inappropriate practices" such as the cultivation of maize on sloping land in dry areas. The peasantry seem embarked on a process of destruction that the government must prevent.

The next causal element, according to this model, is legislation. While there are 14 acts relating to the environment and natural resources, these are almost never implemented and were for the most part enacted by the colonial government. They clearly provide inadequate protection for the natural environment in light of what is happening. Sometimes the policymakers themselves are attacked for their lack of environmental awareness and their failure to play a firmer role in both legislating and implementing legislation.

Within the decision-making structure itself, certain weaknesses are recognized, but, not unnaturally, these are measured against such parameters as

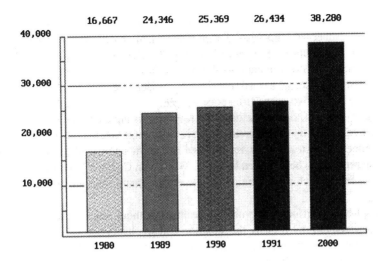

Figure 13.3.   Population of Kenya (in thousands). Annual population growth: 4.2%; population doubling time: 17 years; population density: 42 inhabitants per square kilometer; urbanization: 19.7%. (Courtesy of PC Globe, Inc., Tempe, Arizona, U.S.A. Copyright 1991.)

effectiveness and efficiency rather than general relevance. Typically the sorts of difficulties identified are lack of coordination, the sectoral approach, overbureaucratization, slowness, etc. The principal dysfunction is seen as that between the large and comprehensive nature of the problem and the disaggregated nature of the administrative system (examined in detail in Chapter 12).

Given this assemblage of "causes", it is now logical to proceed to a series of policy measures and administrative "reforms" to deal with the environmental crisis. The "Rabbit Syndrome" leads to a program of intensified family planning and population control education to enable the people to learn how they might control numbers. Undeniably, fewer children added to the population will reduce the pressure buildup on resources in the future. It is a classic technocratic approach to the problem, but we shall see why it is unlikely to work without draconian central enforcement of the Chinese type.

To tackle ignorance, tradition, and cultural attitudes, the natural solution within this paradigm appears to lie with education: teaching people conservation-farming methods and instilling into them an awareness of the damage that they are doing, if they are not already aware of it. There is a need to sweep away the legacy of the past because the damage is being done largely by small, traditional, peasant family farms. Films, television, and radio may be used to disseminate the message.

Inadequate legislation is, perhaps, the easiest aspect to deal with in the technocratic paradigm. All it requires is the will to draft new and tougher laws and to ensure that the institutions and resources are put in place to see that these laws are rigorously applied. On the occasions where these laws have been opposed in the National Assembly, then it is said that the politicians need educating because they do not understand the nature of the problem, or are serving the short-term interests of their constituents rather than the long-term interest of the country. The "need" for new legislation led to Kenya's 1980 Environmental Enhancement and Protection Bill. The word "protection" is significant because, in the context of this model, the government ends up protecting the environment from poor people.

In terms of institutional reform, the response to the problem has been the creation of a ministry of the environment in an attempt to bring the fight against environmental destruction under one roof and show an official determination that the problem is being faced squarely. We saw in Chapter 12 that it is much easier to create such an institution than to ensure that it does anything significant.

Despite these efforts the problem continues without relief. In other words, this approach did not remedy the situation and so clearly something is wrong with it. Before examining the consequences of the technocratic approach, it is useful to look at an alternative paradigm to explain the same environmental breakdown.

The Access-to-Resources Approach to the Problem

In the access-to-resources approach the political economy becomes a variable alongside all the technical parameters previously outlined. In the case of Kenya, this allows us a new level of explanation that will relegate most of the "causal elements" of the technocratic approach to the status of dependent variables. The problem is not a physical one; only the symptoms are physical. It should be stressed that the elaboration of the second approach is specific to Kenya, though is one that could be universally applied. This has been demonstrated effectively for Nepal by Blaikie.[2] The access-to-resources approach requires us to consider the historical evolution of relations between people and the land, and between people and their political and economic systems.

Kenya has gone through not only the colonial experience involving the sudden intrusion of Europeans and European values, but has also experienced a particular form of colonial economy — the settler system. The country was divided spatially, politically, and economically into two distinct camps: the commercial, large-scale farming enterprises and the intensively settled native peasant-farming "reserves" with their subsistence agriculture. This highly divided system made the transition to independence with little structural change other than the fact that the ownership of many of the bigger farms passed into the hands of well-placed Africans. As a result, a different problem arose: grossly unequal access to land, particularly land in the better ecological zones, in a context where there are almost no forms of security or income other than farming. In addition, the Kenya model of development, like many others, is based on export crops and the large-scale commercial farming of food crops for the urban market, forming a barrier to any real modification of the factors causing pressure on parts of the land.

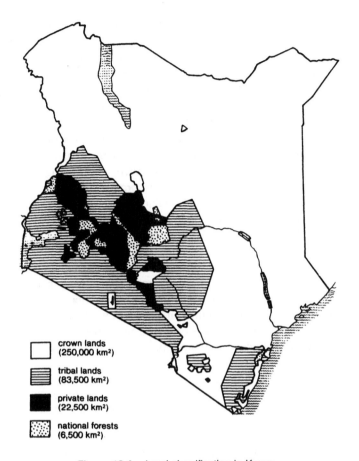

crown lands
(250,000 km²)

tribal lands
(83,500 km²)

private lands
(22,500 km²)

national forests
(6,500 km²)

Figure 13.4. Land classification in Kenya.

During colonial times, as shown on Figure 13.4, Kenya was rigidly divided into the "Scheduled Areas" where only the whites could farm and the "Native Reserves" where Africans were permitted to farm. Only the Europeans were allowed to partake in the cultivation of the principal cash crop, coffee. The Reserves, meanwhile, were regarded as a steady source of cheap labor. People living there could not earn an income by participating directly in the cash crop economy as farmers. They had to work instead as laborers on the European-owned farms. The allocation of 14,000 mi², virtually all of it in the best ecological zones, to 3500 Europeans, and only 52,000 mi² of land to more than 4 million Africans (Figure 13.4), set the scene for the troubles that followed. As population in the Reserves grew, there was nowhere for the surplus population to go since one quarter of all the best land was now denied to them. At the same time, vast areas of dryland grazing were enclosed by European ranchers, confining traditional herders such as the Maasai to a much smaller area and disrupting their well-established grazing circuits. The introduction of basic health-care facilities by the colonial government allowed the population growth rate to begin its rise to today's level; it now has the fastest growth rate in the world (Figure 13.3). By the 1940s the government became seriously concerned about soil erosion in the Reserves. This the authorities attributed to overpopulation, though the overpopulation was, of course, relative to the amount of land upon which the Africans were allowed to live.

The colonial authorities were reluctant to allow the Africans to cultivate maize for sale because this would increase the pressures on the land. To Africans, this reduced the options open for any sort of alternative income other than to work on the European farms. The destruction of the rangelands and the Reserves led the Agriculture Department to propose two main sets of measures: compulsory soil conservation terracing and compulsory destocking in the pastoral areas. Both measures met with fierce opposition from the people concerned. Cultivators felt that the first measure forced them to work harder and avoided the real answer: the return of the lands taken from them by the Europeans. The second measure affected the pastoralists, already facing a declining environment, who saw the government threatening their security by taking away some of their cattle. The government attributed these negative reactions to a "lack of understanding" on the part of the Africans (environmental unawareness?). From this time forward the government saw itself locked in a struggle with a destructive and uncomprehending population:

> Unless some pressure is applied to urge improved methods and practices, and unless such pressure is continuously applied . . . it will not be possible to save the fertile areas in Kenya from deterioration . . . without compulsion under legislation to enforce improved agricultural practices.[3]

The anger over confinement in the Reserves exploded between 1952 and 1955 in the *Mau Mau* uprising. During this period there was widespread violence, particularly in the Kikuyu tribal territories where the land grabbing by the colonial government had been most blatant. The ultimate failure of *Mau Mau* when faced by a European army with modern weapons strengthened the control of the government, which set about creating a stable, loyal corpus of "yeoman farmers" among the Africans. This action did not begin to address the growing problem of landlessness.

With independence (1963), many thought the picture would change. However, the Kenyan government depended on the large farms to feed the urban population and grow the commodities for export that generated 80% of the country's foreign exchange. Using British-provided funds, the Kenyan authorities bought out many of the white farmers, but then replaced them with prosperous black farmers. Leys has noted that "the policies pursued in the 1960s ensured that . . . there would be a structure of agrarian interests and an institutional apparatus strong enough to resist pressures for radical change."[4] The bimodal distribution of land continued with about 7% of the holdings consisting of more than 1800 acres and 30% of less than 2.5 acres. About

PROBLEM	SYMPTOMS	CONSEQUENCES	SOLUTIONS	CONSEQUENCES
The political economy of Kenya	Unequal access to land	Environmental degradation	Radical change	Greater self-reliance and security
	Marginalization of poor	Hopelessness	New forms of security	Diminished need for children
	Lack of alternatives for those short of land	Frustration	Land redistribution	Greater spread of cash economy through greater participation
	The export: crop, urban-based foreign exchange model of development		New priorities in use of land	Diminished resources for urban sector
	Institutional oppression of poor			A stake in the future for rural poor
	Environment over people			
	No security other than land or children			

Figure 13.5. An alternative perception.

99,000 of the holdings were assessed as having no cash income whatsoever. So there were two Kenyas — one inside the world economy and one rapidly falling outside any sort of economy at all.

It is now possible to reconstruct the diagram illustrated as Figure 13.2 into Figure 13.5. We now redefine the "problem" as consisting of the political economy, security, the model of "development", and access to resources. We see a system that accumulates a large number of people on a diminishing resource base, without any alternative form of subsistence. In this case the "environmental crisis" simply becomes an outward expression of this unequal system. The urban areas showed a great appetite for imported goods, energy, and food, all of which had to be paid for by increasing the amount of cash crop exports. This in turn put even greater pressure on retaining the structure of land holding and did not address the problem of the landless and subsistent farmers. In the 1970s the visible trade deficit rose from 64 million to 304 million Kenyan pounds.

In the context of this redefined "problem", access to land is the only form of security open to the poorer population since there are no pensions, urban job opportunities for the rural migrants, or social security. In this scenario of no alternatives, the rural population had *no choice* but to move onto hillslopes, to clear forests, or to expand into the dry zones on the margins of cultivation. It was a case of marginal people moving into marginal lands. Stock keepers attempted to keep more animals to provide some sort of livelihood on diminishing pastures, some of them lost to expanding cultivation, others to prosperous ranchers.

Under these conditions, it is not surprising that people flout the law and that the law is rarely applied. It is quite unrealistic to enforce laws against people who have no alternative, because once the police disappear, the people will return and carry on as before. These same people, perforce, look upon children as security. Whom else can they turn to when they are older and have no way of supporting themselves? A family with several male children may have at least one who will find a secure job and look after the parents. Female children provide a drain on the family's resources because of the marriage dowry system and are therefore less desirable. Since parents cannot select the sex of their children, their desire for security explains the fact that the average Kenyan woman has eight children. Large families are therefore not the result of ignorance about family planning, which offers nothing in the way of security. What results from this situation is a condition of individual rationality: "It is perfectly logical for me to have as large a family as possible." Unfortunately, the inevitable fact is that all these individually rational decisions add up to a situation of collective madness. Still, what is the individual to do?

It is unlikely that the process of education and the raising of environmental awareness have anything particularly immediate or relevant to offer once the problem has been redefined. We may safely assume that the people destroying the marginal lands and the former Reserves know perfectly well what the outcome of their actions will be. They simply do not see any alternatives, and they are likely to be angered by the actions of the government telling them to invest more time and energy into conservation practices. This is especially true when they see the thousands of acres still held under low-intensity cash cropping by the new class of indigenous large farmers. It is simply not fair to speak of the peasantry as ignorant, traditional, and perverse under these circumstances. They are behaving in a perfectly understandable manner, even though it will lead to the rapid destruction of the country's natural resource base.

It is perfectly evident also that passing tougher legislation is not going to do much to stop the degradation. The problem with the law is that it is a negative force. It tells people what they should not do and what will happen to them if they do it. That is all well and good where one is dealing with people who choose to be antisocial when they could be gainfully employed doing something socially acceptable. It is totally meaningless, however, when applied to people who cannot do anything else. The law then becomes an instrument of oppression. In the context of a highly divided society such as Kenya's, it becomes an instrument for polarization, by which those that have tighten their control over those who have nothing. The police recognize this hopelessness, and this is probably why they tend not to enforce the laws. In addition it would be a brave constable who would confront a group of Maasai warriors and serve them with a destocking order. But before we construct this whole argument around conspiracy theory, it is necessary to remember that many of the people cultivating the big farms in Kenya see themselves as the last great hope. For without the products of their farms, there would be no food for the workers in Nairobi and there would be no hard currency to keep the cars, trucks, and power plants running. Any change in the order of things is, therefore, perceived as threatening to the whole order and stability of government.

This is a rather all-embracing form of policy trap. The problem is that the maintenance of the present system and structure allows a growing number of Kenyans to fall out of the bottom of that system, while the marginal and peasant-held land plunges inexorably toward destruction. It goes without saying that tinkering with the organization of the administration by creating, for instance, a ministry of environment, is unlikely to achieve anything at all unless it expands its brief to include poverty as an environmental variable. The adherence of the ministry to the concept of environmental *protection* suggests that its orientation is toward Figure 13.2 rather than Figure 13.5. In short, the various elements of the technocratic approach serve the status quo, do not address the central problem, and provide palliatives rather than answers. Thus the situation continues, and will continue, to get worse. A government cannot hold the environment in trust for future generations at the expense of the basic livelihood of a large proportion of the current population.

The question now arises, "How could we save the environment in Kenya?" The first point to grasp is that, for the moment at least, Kenya is not overpopulated as long as there are around 12 to 16 acres of fertile land per person. The difficulty is that most of the population has no access to this land. But, if they had, what would be the effect? The government's principal concern is that they would tear out the cash crops and expand the subsistence sector, leading to difficulties both in feeding the urban dwellers and in producing commodities for export. On the other hand, if people were brought into the commercial farming areas they could do so on the basis that they continue at least part of the cash crop base. The scale economies might be preserved through cooperative farming arrangements. The regular source of cash income would provide each family with an element of security that would obviate the *need* to have so many children. This, in turn, would reduce the rate of future pressure increase on the land. There is no doubt that the typical economic or agricultural planner could set out a whole slate of problems as to why this would not work. In addition it would require an enormous political leap of faith, which is somewhat unlikely while so many of the big farmers are also politicians and civil servants. What has to be asked at this point is: "Is there really any alternative?"

The industrial and service sectors simply do not have the capacity to grow and absorb the additional population despite the somewhat desperate measures of the government to increase the employment of Africans in these areas by legislation that forces employers to take on more workers. The larger farms will tend to become more capital-intensive in the manner of farms elsewhere, and this does not offer the prospect of increased employment either. There is little prospect for serious diversification of the economy to produce new and expanding job-creating sectors. Tourism is notorious for the leakage of income out of the country and for the very small amount of retained income and job creation. Those landless who drift into the towns will continue to feed the growing rate of crime and instability. To cope with this the government has shown a tendency simply to retrench and acquire increasing power to control at the expense of democratic rights. This was especially true after the failed coup of 1983. At the time this book was being completed, the abuse of human rights in Kenya has reached the point where Western governments have come together to withhold aid until the basic rights of the country's citizens are restored.

This focus on the Kenya case has not been an attempt to show Kenya in a particularly bad light vis à vis the rest of the world. Instead it shows what happens as rural populations grow in the postcolonial structure of heavy dependence on cash crop exports and little else. It seems almost bizarre when we consider that the best land is reserved to grow coffee and tea for the rich countries, while armies of peasants are being forced into a position where they cannot provide even for their own subsistence. In the West, armies of people were displaced from the land during the 18th and 19th centuries, and although they endured periods of grinding hardship, their salvation was to be found in the rapid expansion of employment in the industrial sector and in huge migrations to new lands. In most of the tropical developing world, industrial growth is not a viable market for surplus population and there are no "new lands" anymore. The specter of Malthus appears on the horizon.

REFERENCES

1. *Daily Nation,* Nairobi (November 19, 1980.)
2. Blaikie, P. M. *The Political Economy of Soil Erosion in Developing Countries* (London: Longman, 1985).
3. Clayton, E. *Agrarian Development in Peasant Economies* (Oxford: Pergamon Press, Inc., 1964).
4. Leys, C. *Underdevelopment in Kenya* (London: Heinemann, 1975).

OTHER USEFUL READING

• Baker, R. "Environmental Degradation in Kenya: Two Conflicting 'Explanations'," *Afr. Environ.* 5 (3):15-43 (1986).
• Tierney, J. "State of the Species: Fanisi's Choice," *Science* 7(1):26 (1986).

CHAPTER 14

The Destruction of the Rain Forest:
Development in Action?

The Rain Forest Domain

Closely associated with the humid tropics, the tropical rain forests cover about 10% of the surface of the globe and are concentrated in three main areas: the archipelago of Indonesia (Figure 14.1) and the two huge basins of the Congo (Figure 14.2) and the Amazon. There are smaller outliers, such as in parts of Central America, but about half of the surviving tropical rain forest in the world is to be found in one place — the Amazon (Figure 14.3). This one vast ecological region covers about 1.9 million square miles out of the global tropical rain forest area of 4.06 million square miles. In general this ecosystem is associated with relatively low densities of human population averaging between 0 and 30 persons per square mile, which is the explanation of the forest's survival so far.

In contrast to its low population density, the tropical rain forest is characterized by enormous species diversity and a great weight of biotic matter per acre. In discussing the humid tropical forest biome, mention should also be made of the tropical mangrove forests that fringe the coasts of many parts of the tropics. These forests form a unique habitat, which, in a fragile strip, separates the sea from the land, supports its own life forms, and maintains the stability of the coastline. The tropical rain forest is a world of intense competition for light, for height, and for scarce nutrients in the thin and poor soils. The wealth of the rain forest is *in* the rain forest, not beneath it in the soil. For the governments and developers of the humid tropics, and for those who advise them, this reality has been one of the hardest lessons to learn.

Figure 14.1.   Tropical rain forest in Asia. (Modified from map supplied by PC Globe, Inc., Tempe, Arizona, U.S.A.)

Present Rain Forest Extent
Former Rain Forest Extent

Why is the Tropical Rain Forest Ecosystem Important?

At first glance the tropical rain forest is an overwhelming confusion of plants and animals — vast towering boles and closed canopies. But in Western economic terms there does not seem to be much going on in them and what "development" there is occurs almost exclusively by eliminating the forest. Just as it was in Western history, the forest is seen as a challenge to be overcome, conquered, and cleared away. Darkness is replaced by the intrusion of sunlight. Western folklore is full of the image of the forest as a dark, dank, and dangerous place full of creatures to be feared, suffused by the supernatural, and replete with lurking danger (read *Little Red Riding Hood* for a good introduction to the mythology of the forest). Druids excepted, it was the very manifestation of evil. The development of the West is the story of the death of the forests, first in Europe, and then in a mighty sweep across the United States. Not so strange then that Western development concepts have given the forest short shrift, and have generally taken forest clearance as the first step in the development process. We saw in Chapter 6 how the indigenous people of the tropical rain forest had developed a sustainable living from the forest by encouraging natural regeneration (shifting cultivation). They knew the limits of the soil in this biome. The Europeans *assumed* the same benefits would follow tropical rain forest clearance that had been manifested in the temperate

Present Rain Forest Extent
Former Rain Forest Extent

Figure 14.2.   Tropical rain forest in Africa. (Modified from map suplied by PC Globe, Inc., Tempe, Arizona, U.S.A.)

zone of the north and never asked *why* the forest was not cleared in the preceding 10,000 years.

The tropical rain forests are places where few people live, where few commercial products grow, and where opportunities to make money depend on the destruction of the forest through logging, mining, flooding, or burning. That may be the conventional picture, but the truth is somewhat different. As a natural environment, the tropical rain forest is unrivaled. Half of the planet's plant species live there, and about 1% of them have been studied so far. However, the enormous size and weight of many of the species means that about two thirds of the planet's plant biomass is represented by this one ecosystem. We really have no idea how many insect species are to be found, since most of them remain, at this point, unidentified. We do know that scientists have found more species on, and around, one tree in Amazonia than have been classified in the British Isles. About one fifth of all the bird species of the world are in the Amazon rain forest, and the single African island of Madagascar holds one quarter of all of Africa's known plant types. The richest diversity of birds anywhere on earth is in Amazonian Colombia, while Indonesia has the highest diversity count of mammals anywhere. And that is based on what we know *so far* about this least-taxonomically studied region.

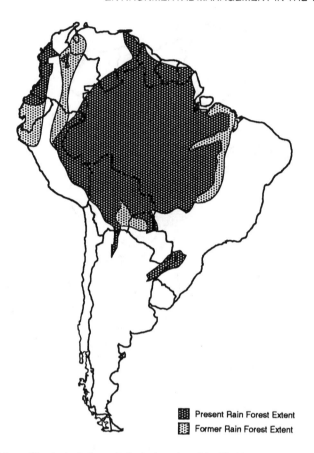

Figure 14.3.   Tropical rain forest in Latin America. (Modified from map supplied by PC
Globe, Inc., Tempe, Arizona, U.S.A.)

Bearing in mind the last point — that the vast majority of plant and many
faunal species remain unidentified and unstudied — it is salutary to note that
already over one quarter of the drugs prescribed in the United States come from
tropical plants, including the tropical yam, which yielded the key to the birth
control pill. The prospects for products of value from the more than 90% of the
plants and animals remaining to be researched are clearly very encouraging. A
summary of the benefits we can and might derive from the tropical rain forest
is given in Table 14.1.

The value of the tropical rain forest can be measured in other ways as well.
This vegetation/faunal climax is the environmental end product of the tropical
humid climate, which is characterized by enormous precipitation totals, often
received in dramatic, heavy thundershowers. The impact of the daily deluge is
broken by the almost closed canopy atop the trees. Gradually the water,
dissipated of its destructive force, makes its way to the forest floor, seeps into
the groundwater, and flows steadily into the rivers and back into the vegetation.

Table 14.1. The Benefits of the Tropical Rain Forest

Subsistence needs	Environmental uses	Industrial uses	Genetic store
Fuelwood	Preventing soil erosion	Poles	Strains for crops
Charcoal	Watershed protection	Plywood	Medicines
Building materials	Soil fertility	Veneers	Industrial chemicals
Fodder	Shade	Papers and boards	
Fruit	Shelter from wind	Gums and resins	
Nuts	Flood prevention	Oils	
Honey	Landslide prevention	Exports	
Medicines	Water retention		
Dyes			

Source: "The Disappearing Forests," UNEP Environmental Brief No. 3 (Nairobi, Kenya: United Nations Environmental Program).

The ecosystem functions as a sponge, allowing steady penetration and flow of otherwise deeply destructive forces in and out of the environment. If the forest were not there, the water would strike the ground directly and would soon result in dramatic soil erosion. The increased rate of runoff and diminished groundwater recharge would lead to serious and violent overloading of the rivers, the building up of the river bed through deposition, and, eventually, destructive flooding. The 6000 deaths from flooding and landslides on the Philippine island of Leyte in late 1991 were out of all proportion to the size of the tropical storm, which had winds of only 40 to 60 mph. However, deforestation increased the flood potential in the catchment flow to six times over normal and that is why so many people died. Massive deforestation of the foothills of the Himalayas in Nepal has caused huge inundations in the lower Ganges, causing the death of 10,000 people in 1978 when the flow reached 40 times the "normal" level. The soil washout from this forest destruction is actually constructing a new island in the Bay of Bengal. Unfortunately the rival claims to this island come from Bangladesh and India, while Nepal — which donated this new piece of the earth — has no claim since it has no coastline. Such changes in the sediment load, velocity, and volume of major rivers would also threaten to overwhelm the sensitive environment of the mangroves as well. To get some idea of the forces at work it is instructive to bear in mind that the Amazon, right now, would fill Lake Ontario in *3 hours*. The idea of such a force out of control is hard to imagine.

There is also some evidence, though not what we might yet term "proof", that the tropical rain forest is a significant factor in influencing global climate. Trees are a repository of carbon, which they take in as carbon dioxide in a process that yields a significant part of the oxygen in the world. In a time of global warming, regarding which CO_2 is a significant culprit, it is to our benefit to preserve this huge carbon "sink". Conversely, burning the forest releases the CO_2, and eliminates part of the machinery for processing carbon dioxide into the oxygen we need. Locally, the vast amount of moisture transpired into the regional atmosphere by the trees seems to be a key element in maintaining the integrity of the ecosystem. In central Panama, for instance, where most of the forest has been stripped, a steady decline in local rainfall totals has been measured. This reduction is not matched at the coastal extremities, and so the downward trend cannot be blamed on a wider climatic shift. To be frank, we do not know the exact role played by the tropical rain forests in terms of global climate, but given their extent and weight of biomass, it could well be significant. Then, again, we do not know the true role of the other huge player, the oceans. This should not, however, lead us down the pathways of recklessness.

The point that needs to be stressed is that the tropical rain forest is an "old growth" ecosystem. Because of the enormous forces at work in terms of total precipitation, temperatures, and dynamic energy, the tropical rain forest does not benefit from interference. It has almost no intrinsic store of value other than in the forest itself — the soil is almost worthless. Its stability is maintained by the forest, its mesoclimate is fueled by the forest, its groundwater regime is created by the intervention of the forest, and so on. This is very much *the world of the forest*. Let Western beings view the forest as a challenge, and an obstruction, at their peril. The indigenous land-use systems of harvesting and regenerating typically fell in with the ways of the forest. However, sustainability is threatened now, and what took Europe centuries and the United States 200 years is happening now in the tropical rain forest in mere decades.

The Empire of the Chainsaws

The destruction of the rain forest is outlined in Table 14.2. For the period covering the last 100 years, estimates of rain forest destruction run as high as 4 million square miles. Currently, across the globe, the figure for destruction averages about 2% annually, resulting in the loss of an area about the size of Kansas each year. These figures are always somewhat speculative, and recently the World Bank has suggested that the total recordable loss of rain forest in the Amazon is around 8% overall and that the rate of destruction has slowed. What is more alarming is the recent nature of much of this destruction, perhaps nowhere better exemplified than in Central America. In that part of the world, about two thirds of the forest has been lost in the last 30 years, so inhabitants there have, in a little over a generation, seen the transformation of the characteristic landscape. In Indonesia estimates run at around 2.25 million acres of destruction annually. In a 1989 study done for the World Bank, Mahar demonstrates the accelerating nature of rain forest destruction worldwide:[1]

- 1975 — 18,750 mi^2
- 1980 — 78,120 mi^2
- 1988 — 375,000 mi^2

These figures, if accurate, highlight a disturbing trend, namely an *increasing rate* of destruction. To gain some idea of where this trend could lead, one need go only to Haiti, where relentless forest clearance has left an exhausted, eroded, and almost worthless landscape.

Table 14.2. Rain Forest Destruction Worldwide

Area of closed tropical forest cleared annually (thousands of hectares)		Percentage of national forest cleared annually, 1975–88	
Brazil*	8,000 (c. 19,500 ac)	Costa Rica	7.6
India	1,500 (3,750 ac)	India	4.1
Indonesia	900 (2,250 ac)	Thailand	2.5
Myanmar	677 (1,692 ac)	Brazil*	2.2
Thailand	397 (992 ac)	Myanmar	2.1
Vietnam	173 (432 ac)	Vietnam	2.0
Philippines	143 (357 ac)	Philippines	1.5
Costa Rica	124 (310 ac)	Indonesia	0.8
Cameroon	100 (250 ac)	Cameroon	0.6

*1987 only.

Source: World Resources Institute. Tropical Forests: A Call to Action (Washington, D.C.: World Resources Institute, 1985).

Who is Doing What to Whom?

The next task is to determine *why* rain forest destruction is happening. Despite the caricature of the "villain" of the tropical rain forest, it is not the logger, but the indigenous peasantry that is often responsible for most of the destruction. Regionally, shifting cultivation running out of control is responsible for 70% of forest destruction in Africa, 50% in Asia, and around 35% in the Americas. Strictly speaking, this is not just a matter of uncontrolled human growth in a system with a low critical population density. In the cases of Brazil and Indonesia, many of the peasants are being encouraged into the forest by governments seeking to solve problems of population pressure elsewhere (such as in Brazil's northeast) which has already been ravaged and destroyed. If, as in Brazil, uncontrolled population growth tends to end up in an urban context, it has the potential to be politically explosive, and an easy way out is to give potential troublemakers land grants in the forests along new roads, often those built by loggers. The critical point to note here is that the land is being given for settlement, which implies *permanence* (continual cultivation of the same plot), and this permanence is completely contrary to the mobile and shifting concepts applied to the forest by traditional farming methods. Writing in *World Development* in 1991, Southgate et al. demonstrate how, in the forested lowlands of Ecuador, peasants clear land in order to gain secure tenure rights from the government.[2] The tenure process is desperately slow, and the peasants must keep the land cleared to demonstrate "use". During this time it is subject to continual pressure. Related to this aspect of the tenure problem, the same writers found that the traditional practice of forest fallow among the Amerindians was abandoned after government land agents informed the indigenous communities that fallow land could be claimed by colonists as "not in use". Property rights, in these conditions, *require* continuous clearance.

It is this element of permanence that is the truly destructive element in the equation. Settled farming cannot be attempted realistically without replacing the source of soil nutrients and those disappeared when the forests went. Livestock or maize will not put them back, and few of the farmers have the resources initially to buy fertilizers. This is what breaks the back of the system. Contrary to what is popularly portrayed, most of the forest is taken over by *small* farmers and ranchers, not logging or livestock multinationals.

The question of whether the farming activity that follows deforestation in the tropics is sustainable or not does not normally enter the equation, which is, instead, made up of economic and political expediency and the search for a short-term palliative. This is the story in Latin America and Indonesia. The same, for the moment at least, does not hold true in Zaïre, where the total population density is extremely small and so unable to effect massive change for the moment. Sadly, as we saw in Chapter 6, the cleared forest soil is not able to support agriculture for more than a few years, then nutrients are drained away, the vegetation that recycled them is gone, and everything downgrades to the exhaustion point. It is not without reason that peasants in Panama are called *arreras* after the local leaf-cutter ants that strip out whole areas of greenery.

Indonesia had a plan to move 1 million of its people from the heavily overpopulated hill slopes of Java to the outer island of Sumatra, which is both forested and thinly populated. Java, however, is a volcanic environment with rich and stable soils cultivated for centuries under a system of terraced irrigation; Sumatra is not. The results were predictable, leaving migrants hopelessly ill-equipped to deal with an unfamiliar environment and the rapid rundown of their new farms. The idea has been abandoned for the moment. When those in the West say that both Brazil and Indonesia are buying off a political or demographic crisis by mortgaging the future of the rain forest, that is fine and fair, and probably accurate. On the other hand, it is the responsibility of such accusers to suggest a few workable alternatives, which we shall attempt to do later.

There is another form of farming, however, that has no traditional roots in the rain forest, but which has gained a destructive foothold there: livestock raising for beef production. It seems remarkable that a land-extensive activity usually associated with drier regions should be taking over one of the wettest places on earth. It sounds absurd, and interestingly, it is.

Between 1966 and 1978 Brazil set out to become the *world's* leading beef exporter by 1980. To achieve this target, 31,000 mi^2 of forest were cleared. The government set up a series of incentive programs, consisting mostly of tax breaks and subsidies, to fulfill this ambition. Corporations moving into the forest and clearing the land were given tax credits of 50%. Over two thirds of the corporations that moved into the forest were livestock companies. The attraction of this venture to the Brazilians was that meat is a foreign exchange earner, and this is a factor that has become more, not less, critical since the Latin American debt crisis broke on the world in 1982. Needless to say, the rain forest is hardly the place to raise cattle. After the initial flush of nutrients is used up, and the inherent sterility of the soil begins to assert itself, the realities creep in. Brazilian livestock raised under these conditions yields about 45 lb of beef per acre, compared with European totals of 535 lb per acre. Furthermore, it takes about 15 acres to support each animal, and so the demands this system puts upon the land are enormous. As the land exhausts and erodes itself, more land is swallowed up from the "free good" of the neighborhood forest. However, in a world of hyperinflation — which was characteristic of Latin America for much of the period since the last war — land is a valuable hedge against the runaway collapse of currencies, especially when the government subsidizes farmers to acquire and abuse the land.

By 1980 Brazil, for all its hopes, remained an importer of beef. In Costa Rica, where livestock fever had reached epidemic proportions, exports doubled while the domestic consumption of meat actually fell, and most of the forest disappeared. Indeed, the average American cat consumed more beef during this period than the average Costa Rican. The great attraction of the system, and there was one, was that it produced *cheap exportable* beef, for which there is, in the United States, an almost insatiable demand. The forests become, by this process, locked into the fast-food market in the United States, a country that still has the capacity to startle foreigners with the extraordinary cheapness of both its beef and its gasoline. Both these bargains have been achieved at a terrible cost. So, before we point accusing fingers at venal loggers and rapacious ranchers, the real agent of doom for the rain forest may be the hamburger worshipper.

Logging the old growth rain forest using conventional practices is quite extraordinarily wasteful and destructive. As has been noted, the tropical rain forest is typified by enormous species diversity and is home to some of the most precious, slow-growing hardwoods in the world. This diversity means that loggers will not find pure stands of what they seek, but must extract their product tree by tree from some of the most inaccessible places on earth. About 2% of what grows in the rain forest is of commercial interest to the logger. In extracting that 2%, however, about 10 trees are damaged for each one extracted, and some 30% of the timber is left on the forest floor. In Malaysia, for instance, even after the tree reaches the mill, only about 40% of it is commercially retrieved. This compares with 98% in mills in Sweden. Globally, after logging is completed in most tropical rain forests, about 55% of the logged land is subsequently deforested completely.

The reason for the subsequent clearing is an indirect consequence of logging, since that industry cannot function without roads. These roads are the arteries along which the landless and the speculators subsequently move to carve their clearings out of the forest. This makes each logging venture merely the starting point of a much larger and more destructive wave of activity.

Several forces encourage the continuation of logging as essentially a *mining* of resources (extraction with no thought for the future). First, the tropical hardwoods are a valuable export commodity that, for a country such as Brazil owing $117 *billion* and having unmet interest payments alone of $8 billion, is too attractive to ignore. The slow-growing nature of these hardwoods means that no logging enterprise is interested in replanting the same species. The payoff would be so far in the future, in purely financial terms, that the investment could never be justified in the economics of conventional costs and benefits.

Second, nearly all of the forests are on state-owned land, so public policy is extremely critical in determining their use. The practice has been to offer these lands on short-term concessions (they are not bid for at auction like oil concessions for some reason). This short-term perspective accentuates the extractive, destructive nature of the logging industry's approach. Furthermore, as is commonly the case in Asia, many of the companies represent foreign interests that have no stake in the sustainability of the host country — especially as the forests seem endless and inexhaustible.

Third, the concessions are *valuable* and, alas, are a source of illegal income for some and of extreme political pressure for others. It is hard to imagine how serious reform of the logging industry is going to occur in the Philippines, for example, when loggers are the largest contributors to party funds in that country. Overall, the mining of the forests represents a quick-fix approach to development. Even in the countries with the largest exploitation of rain forest resources for timber, the work force engaged in this enterprise *never* exceeds 1% of total employment. The logging industry is, perhaps, the best (worst?) example of the state's failing in its role of stewardship of natural resources and of the state's desire to spend capital as income.

Finally the forest has also been consumed by hydroelectric schemes and mining ventures, though neither of these compare with the activities already discussed. Still, in the last decade, Brazil's Tucurui Dam drowned 912 mi^2 of rain forest, and its Grande Carajas iron ore deposit — the largest in the world — requires 3500 mi^2 of forest annually to keep its smelters supplied with charcoal.

Polarizing Forces

Virtually everyone who follows the news will have heard about and formed an opinion about the destruction of the rain forest. This has led to a polarizing and sometimes unrealistic hardening of opinions. Politicians in some countries, particularly Brazil, which has borne the brunt of media onslaught, have become paranoid about the issue — but no more paranoid than many of the people, including rock stars, who are pointing fingers at them. It is necessary at this stage to consider the preconditions for finding a sustainable option to manage this valuable resource. It is perfectly understandable when people in the West say that the rain forest is a "world resource". The fact is that these forests happen to lie within the national boundaries of sovereign states where politicians with hard decisions to make reside. Since rain forest activists are therefore still trapped within the conventions of sovereignty and nationhood, they need to be realistic about the process. Furthermore, if tropical rain forests are a world resource, then the world is going to have to do something concrete about managing them — which is another way of saying helping politicians to face up to the terrible choices they must make, including the provision of resources to allow them a realistic range of ecologically sustainable options in their choices. Driving the politicians further into a corner does no good at all. Guilt will not solve the rain forest dilemma; realism might.

It needs to be understood that many of the countries in question have a race in progress between the short-term constraints of their debt situation and the long-term requirements imposed by sustainability. A debt overhang does not lead to conservationist policies; it leads to short-term asset stripping, as we have discussed in Chapter 10. The West cannot with one voice say "stop destroying *our* heritage" and with the other say "have you got the $8 billion you owe us?" We must face these aspects as two sides of the *same* coin. It is in this situation of paradox that polarization develops, which is a totally unhelpful environment. We have the experience of former President José Sarney of Brazil saying "the foreign media are promoting an alarmist campaign against Brazil directed at making slaves of the Brazilians."

His successor, President Collar de Mello, said: "Brazil cannot be put in the defendant's dock as the cause of today's environmental problems." Very significantly, he followed that statement by saying that he *is* concerned about the environment and that, in fact, "it is *second* only to debt as one of the world's two great problems." The rank order is significant: short-term first, long-term second. But what politician can afford to think otherwise in such a situation? Right now taxes and subsidies exaggerate the process of destruction, while the long-term loss of resources and the undermining of sustainability do not show up in the cost/benefit equations. Environmental destruction has become a political palliative and a workable tax shelter.

What Can Be Done?

First, we must get rid of the conspiracy mindset. Few people, probably none, are trying to destroy the rain forest as an end in itself. They are trying instead to make a living, make a profit, stay alive politically, or feed their family. That is the context. Is there a sustainable answer?

In many countries land is scarce because of its distribution, as in Kenya. The Kenya situation of most of the best land being in the hands of the few has been the norm for centuries in Latin America, where even now 93% of arable land is in the hands of 7% of the population. Of course, much the same could be said for the United States, but the majority of the U.S. population *does not depend on agriculture for its livelihood*. What we need is a lasting way of supporting the 500 million people who inhabit rain forest areas.

Work by the New York Botanical Garden's Institute of Economic Botany has shown that *managing* the forests on a sustained-yield basis would produce returns twice those of cattle raising or lumbering. The botanical garden study showed that a harvest of fruit, latex, and timber together would be worth

Text Box 14.1

Some Brazilian Perceptions of What is Happening to Them

The United Nations Environment Program, from time to time, commissions opinion polls to discover how different countries perceive environmental issues. In 1989 Brazil was added to the list, with some interesting divergences from the general pattern of opinion. The poll surveys the "people in the street" and a group identified as "leaders." The following are the positive responses to some questions:

Q.1: Do you agree that environmental problems are a domestic concern only?

	Public	Leaders
All countries	34%	30%
Region (C. & S. America)	44%	37%
Brazil	64%	8%

Q.2: Do you agree that more should be done to protect the environment?

	Public	Leaders
All countries	90%	92%
Region	92%	95%
Brazil	92%	96%

Q.3: Do you agree that protecting the environment should be a major priority of the government?

	Public	Leaders
All countries	85%	85%
Region	85%	78%
Brazil	65%	6%

Q.4: Do you agree that life in this country is so difficult that the environment is not a top concern?

	Public	Leaders
All countries	39%	38%
Region	48%	43%
Brazil	58%	52%

Source: United Nations Environment Program. *Our Planet* 1(4):11 (1989).

$9000, compared with $3000 for livestock — and the forest harvest could be sustained, which is a doubtful proposition for livestock. Only 10% of the $9000 was for the timber. Indeed the equation becomes even more attractive when the subsidies and tax breaks are taken away from the livestock enterprises.

Charles Peters et al., writing in *Nature* in June 1989, showed how researchers sampled a 1-km^2 site at a village south of Iquitos, Peru, and noted the existence of edible fruits, rubber, commercial timber, medicinal plants, and other forest products.[3] Taking into account the cost of transportation, labor, and the current market value of the crops, *selective collection* had a better return than simply cutting over the area. Extracting the timber would give a once-and-for-all yield of $1000 net at the mill. Periodic use, including selective cutting, latex collection, and fruit gathering, would bring the value of the area up to $6820 — and returns would be ongoing.

Another innovation is "strip shelterbelt" management, whereby a gap 20 to 50 m wide is cut. This opening triggers the rapid growth of many species because of the intrusion of sunlight (the "light gap" phenomenon) and allows for the preservation of species diversity. By such means, it is suggested, a rotation period of 40 years for sustained production of hardwoods is possible — as opposed to the 100-year figure conventionally adopted.

There is, of course, also the great "unknown" of what lies untapped in the remaining rain forests. Research alone will unlock the potential of new products. This research should incorporate the traditional practices of those who have lived in the forests for centuries. Too often in the past, rain forest dwellers have been hunted down or herded like zoo animals. Culture, too, is a resource, and this has been destroyed even faster than the trees. We already have the concept of the **biosphere reserve** under the aegis of the United Nations Education, Science, and Cultural Organization (UNESCO), ever since they initiated the program in 1970. This concept allows for a sufficiently large area to be protected at the core so that the diversity may persist. Around the core are graded buffer zones that gradually introduce different forms of suitable land use. The biosphere reserve allows people the *option* of preserving a resource that natural and social science may then explore.

The main aim of rain forest management must be to persuade the nations of the world to work together to save this massive resource. This does not mean turning these entire regions into parks; doing so would not really address the needs of the countries where the rain forests are found. Governor Gilberto Mestrinho of Amazonas, Brazil's largest state, said that environmentalists "want to keep the Amazon like a circus, with us as the monkeys." Any plan *must* incorporate use, as well as preservation.

Unfortunately, the countries with rain forests are all developing countries, whose populations have expectations and who do not take lightly to being told to freeze into a preternatural state and not do the one thing that every developed country did on its pathway to riches: cut down trees. People still need land or jobs. If they need land, then they also need a system of cultivation that maintains the quality of the land. Maize monoculture is not the answer. In each of these areas serious research needs to answer the question: How is a peasant, with almost no resources, going to make a living without destroying this piece of land and go on making that living from this piece of land *and* harbor realistic expectations that life can get better for his or her family?" Governments will want to know what, if anything, will replace the earning from logging royalties, the foreign currency income from timber exports, and the dollars from the export of cheap beef. These considerations cannot simply be *eliminated*; they have to be replaced.

Those who hold the debt must work with those who owe the debt. Debt-for-nature swaps are not the answer. These are seen as patronizing and as a sellout by many in the developing world. The term "ecological colonialism" has been used to describe these attempts by Western conservationists to buy discounted debt — mostly in Latin America — and wring out of the governments promises to behave better. We have to work to provide sustainable economic alternatives, not to turn the rain forest into a nature reserve. This means making sustainability economically attractive. If the forests go, then maybe the climate goes with them — bad news for all of us, and something for which we may be willing to pay avoidance money. Moralizers will have to be certain they know exactly *where* that hamburger, that teak salad set, and that mahogany buffet came from.

China, much of which was ravaged by deforestation millennia ago, now gives land titles based on *replanting* forest. Korea has replaced 70% of its area with new forests — land that was effectively deforested during the Korean War. Somehow, we have to work into today's budget the fact that the Panama Canal is silting up because of deforestation. If Western countries want to use the Panama Canal, then they cannot also have cheap beef produced in Panama the way it now is. Sustainability should be at the root of all major investment schemes. Give tax breaks for those who retain the value of the land, not those who mine it. There are ways to use the forest and its resources without first eliminating it, but there is all too little research about this and especially about the practicalities of the economics of such systems. Too often the research looks at the economic value of forest products in places where no market infrastructure exists. Also, the world market for some forest products has dropped dramatically since the encouraging calculations were first done. Latex is a good example.

There are some positive signs, though we must not forget that the overall trend has been toward accelerating destruction. President Collor has made the scrapping of livestock subsidies a major policy plank of his government's policy, though he argues that most of the livestock farmers are too small ever to have received a subsidy. He has also enacted an "extractive reserve" where the ideas of sustainable management will be tried out in an area about the size of former West Germany. Finally, the Brazilian government has proposed the ecological zoning of the whole Amazonian area to govern land use. Part of the problem in a remote area of this size is knowing what is going on and having the resources to back up policies on the ground. With Brazil hosting the 1992 World Environment Conference, it is likely that the tropical rain forest may become the test case for the application of new thinking about sustainability and development in a global environment.

REFERENCES

1. Mahar, D. *Government Policies and Deforestation in Brazil's Amazon Region* (Washington, D.C.: The World Bank, 1989).
2. Southgate, D., R. Sierra, and L. Brown. "The Causes of Tropical Deforestation in Ecuador: A Statistical Analysis." *World Dev.* 19(9):1145-1151 (1991).
3. Peters, C. M., A. H. Gentry, and R. O. Mendelsohn. "Valuation of an Amazonian Rainforest," *Nature* 339(2):655-656 (1989).

OTHER USEFUL READING

- Caufield, C. *In the Rain Forest* (New York: Alfred A. Knopf, 1985).
- Colchester, M., et al. "Banking on Disaster: International Support for Transmigration," *Ecologist* 16(2 and 3):71 (1986).
- Dufour, D. "Use of Tropical Rainforest by Native Amazonians," *Bioscience* 40(9):652-659 (1990).
- Fearnside, P. M. "A Prescription for Slowing Deforestation in Amazonia," *Environment* 31(4):16-24 (1989).
- Gradwohl, J. and R. Greenberg. *Saving the Tropical Forests* (London: Earthscan, 1988).
- Moran, E. *Developing the Amazon* (Bloomington: Indiana University Press, 1981).
- Nations, J. D. and D. I. Comer. "Rain Forests and the Hamburger Society," *Environment* 25(3):12-20(1983).
- Repetto, R. "Deforestation in the Tropics," *Sci. Am.* 262(4):6-42 (1990).
- Sesser, S. "Logging in the Rain Forest," *New Yorker* (May 1990), pp. 42-67.
- Shane, D. R. *Hoofprints on the Forest: Cattle Ranching and the Destruction of Latin America's Tropical Forest* (Philadelphia: Institute for the Study of Human Issues, 1986).
- Stone, R. D. *Dreams of Amazonia* (New York: Viking-Penguin, 1985).
- United Nations Environment Program. *Our Planet* 1(4):11(1989).
- United Nations Environment Program. "The Disappearing Forests," UNEP Environment Brief No. 3. (Nairobi, Kenya: United Nations Environment Program).
- World Resources Institute. *Tropical Forests: A Call to Action* (Washington, D.C.: World Resources Institute, 1985).

CHAPTER 15

Desertification — What is it Really?

Most of us are familiar, if only through popular media coverage, with the situation outlined in Chapter 14 regarding the tropical rain forests. However, we are less conscious of the steady destruction that has been occurring in the dryland areas of the tropics, along the margins of cultivation, in the irrigation basins, and across the vast horizons of rough grazing that support the pastoral communities or, at least, supported them in the past. Where we are conscious of the dryland crisis at all we tend to see only the peaks, not the great mass of the environmental tragedy that has been building for decades. Thus we are all familiar with the catastrophic "droughts" that swept Ethiopia in the 1980s and indeed plagued most of the dryland areas of that continent between 1968 and the present time: absolute situations of ultimate hardship when the life-support mechanism collapses completely and people starve or walk endlessly in pursuit of some last desperate hope of survival.

In these semiarid areas writers have, for several decades now, written about a condition they term "desertification". In its simplest sense this describes an expansion of desert-like conditions beyond the limits that were believed to be reasonably fixed until the 1960s. However, the use of the word "desertification" raises many questions. Are we implying that the deserts are physically expanding, so that the dunes are moving to bury well-established areas of settlement and cultivation? Are we suggesting that the climatic conditions that produce the aridity of deserts are worsening to create greater geographical expanses of dry barrenness across the globe? Are we suggesting that "desertification" is a *phenomenon* or a *process*, or perhaps both? Is the spread of deserts a result of something that people have done recently to change the ways in which they have traditionally related to very limiting ecosystems throughout history?

A closer examination of this whole issue will, unfortunately, raise even more questions than answers because it becomes seriously entangled with fundamental problems concerning what constitutes *proof* and what is the nature of evidence, as well as some fascinating though frustrating confusions of *cause* and *effect*.

The Phenomenon

As an observable condition, desertification may best be described as a negative change in the productive capacity of semiarid areas of both cultivation and livestock-raising, leading to a geographical expansion of desert-like conditions. This may take many forms: the salination of irrigation lands, leaving a crust on the surface and generally poisoning the growing horizons of the soil (currently about as much irrigated land is lost to desertification annually as is added); the replacement of perennial grasses by annual grasses and eventually succulents; a general thinning-out of the number of species, and a growing preponderance of the **xerophytic** — or drought-tolerant — species; the growing incidence and crippling effect of drought on food production at the most basic level (Figure 15.1).

Basically an area of dryland is subject to desertification when some force, or forces, starts to denude the surface of its vegetation cover. At first this may be selective, such as by the overgrazing of certain palatable species by goats, cattle, and other domesticated animals that favor certain plants over others. As the grazing and browsing resources are diminished, there develops even greater pressure from animals, now desperate for food in times of hardship or even during the regular dry season. This asset stripping lays bare the surface of the soil and subjects it to baking by the sun, compacting by hooves, torrential downpours, and the now-unimpeded wind. Gullies form in the areas of hilly terrain. Elsewhere general sheet erosion by wind and water takes place and resources are rapidly lost. In the West African drought of the 1960s, it was estimated that, in one year alone, over 60 million tons of dried-out topsoil was blown out of the Sahel and into the Caribbean via the jet stream. When we consider that it often takes 100 years for an inch of soil to form in these regions, the loss is devastating. As runoff increases with diminishing vegetation cover, the buildup of groundwater — for which the "sponge" effect of vegetation is vital — diminishes. Wells dry up, and animals start to die. The hardened, baked surface becomes inhospitable to seed germination, and the process of degradation starts to feed on itself and accelerate.

It is true that occasionally, as in southern Tunisia for instance, there are some dramatic instances of blowing sand burying fields and homes, but for the most part what may be observed is a general paralysis and decay from within — not a situation of being overwhelmed from without. The patient simply withers and dies, rather than being assaulted or mugged by some outside force.

The use of orbiting satellites has allowed us to monitor the state of the major ecosystems of the world on a regular, reliable, and quantifiable basis (see Table 15.1). This has allowed us to see, most dramatically, the apparently

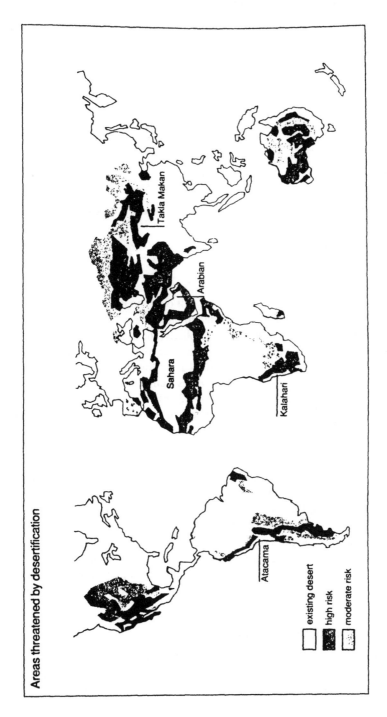

Figure 15.1. Areas threatened by desertification. Existing deserts are named.

Table 15.1. The Regional Status of Desertification Trends

World region	Rangeland	Cropland	Irrigated	Forest	Groundwater
Sudano-Sahelian Africa	•	•	•	•	•
Southern Africa	•	•	•	•	—
North Africa	•	•	•	•	•
Western Asia	•	•	•	•	•
South Asia	•	•	•	•	—
Asian (former) USSR	✓	✓	✓	✓	?
China and Mongolia	✓	•	✓	•	?
Australia	—	—	•	—	—
Mediterranean Europe	—	•	✓	✓	—
South America	•	•	•	•	•
Mexico	—	•	•	•	•
North America	✓	✓	✓	✓	•

Key: • desertification accelerating; — desertification static; ✓ desertification
       improving.

Source: United Nations Environment Program.

inexorable march southward of the uncultivable Sahara into the Sahelian belt
of dryland farming in Africa's Sudanian belt over the last 25 years. In Sudan,
for instance, the desert edge is estimated to have moved south about 60 miles
in the period between 1958 and 1975. Globally, the United Nations Desertification Unit in Nairobi, Kenya, has estimated the total areas subject to desertification, in varying degrees of severity, as follows: waterlogging — 62.5
million acres; salination — 50 million acres; range deterioration — 9 billion
acres; and general drylands exhaustion — 425 million acres. In addition, some
850 million people live in the dryland areas that cover about 35% of the surface
of the earth, and of these, 78 million live in areas currently classified as
undergoing *severe* desertification. Though we tend to think of this as a particularly African problem, in fact larger numbers of people live in severely desertified
lands in both Asia and the Americas. But we should remember that a far greater
proportion of the people most seriously affected are animal raisers in Africa,
contrasted with crop raisers in Asia and the Americas. Some mention should
be made at this point of the fact that the term "desertification" has been used
in a wider sense. One example would be the consequences of clearing the rain
forest and cultivating the cleared land to, and beyond, its point of exhaustion.
In fact this leaves an area that is about as worthless as a desert, but in this
chapter we shall use the term "desertification" to refer to environmental
degradation in *dryland* areas only.

Why is Desertification Important?

At first glance, threatened areas seem to be already extremely poor, especially to the Western eye, and so we may ask why this loss is significant. They
are often vast unimproved rangelands where it may take 100 acres to support
one animal; or they may sustain thin and already risky crops of sorghum. Of

> **Text Box 15.1**
> ## Desertification: Some Facts and Figures[3]
>
> - About 8750 million acres of land — an area the size of North and South America combined — are affected.
> - Every year about 15 million acres of land are irretrievably lost to desertification and a further 52.5 million are so degraded as to become uneconomic for crop production.
> - The rural population affected by serious desertification rose from 57 million people in 1977 to 135 million in 1984.

course, any loss of natural productive resources is a cause for regret as we diminish the value of our heritage. On the other hand there are some special factors to consider here.

The word "marginal" is going to appear extensively in this chapter. These lands are marginal in many ways. They usually lie between the cultivated lands and those lands where cultivation is not possible or is at least far too risky to support viable human communities. The areas of true desert really do not interest us in this chapter, since virtually no form of sustainable agricultural or pastoral land use is possible in the real desert. Indeed, the term Sahel, which is given to the great swathe of seasonal drylands across West Africa south of the Sahara, actually comes from an Arabic word meaning "barrier" or "shoreline", implying the demarcation between two very different worlds. Boundary situations are almost certain to be less stable than the greater, consolidated masses on either side and are consequently very susceptible to change.

The lands are marginal also in the sense that they lie at the limit of acceptable risk. The probability of crop failure, animal death, or drought is high, and societies have had to take that into account in deciding whether and how to live there. In such circumstances a relatively small absolute change, say in precipitation totals or groundwater availability, can spell the difference between viability and nonviability of the system to support the community in absolute terms. In other words, when degradation occurs in more humid areas, it may seriously diminish the options available to the community in that area. In the drylands, such a degradation may simply move the area out of the useful category *completely*, rendering it barren for decades to come. That is what gives desertification its truly terrible importance: it represents a finality, a write-off, an admission of defeat. There are no alternatives after that point.

For the people living in such areas, the alternatives to desertification control are migration, the prospect of terrible suffering as drought becomes famine, and death from diseases that carry off a large proportion of already seriously weakened people. For the policymakers there is the problem of what to do with tens of thousands and sometimes millions of people displaced from their life-support system — people we may call ecological refugees.

Figure 15.2. The Sahelian "boundary" belt. (Modified from a map provided by PC Globe, Inc., Tempe, Arizona, U.S.A.)

To make matters worse, very often the dryland areas of the tropics are also coincidental with some of the poorest communities and nations. A transect through the Sahel region of Africa (Figure 15.2) takes one through countries such as Mauritania, Burkina Faso, Niger, Mali, Ethiopia, the Sudan — some of the poorest, and indeed we might say most "marginal", nations in the world. This means that these countries (1) can least afford to lose any part of their natural resource base, (2) are least able to mobilize the resources to do anything about the situation, and (3) have almost no other options to offer the populations that will be displaced by desertification from their own means of production. So, the impact is doubly great, and the net effect is to take already poor people over the line from poor to totally destitute, and sometimes beyond — to extinction.

Here, then, we have our parallel to the sweeping destruction of the rain forests. It seems as though the tropics is caught in a vice, squeezing out its resources from both ends — humid and semiarid.

How the Problem Manifests Itself

As mentioned earlier, desertification is a creeping paralysis. However, our attention is drawn to the phenomenon not by the inexorable decline but, as is so often the case, by the incidence of *extreme events*. In this case the extreme event that most notably comes to mind is **drought**. One of those reversals of cause and effect referred to earlier occurs when people see desertification as *caused* by drought. It is not. The seriousness of the drought, however, may be a measure of the state of desertification. Drought, by definition, is something exceptional. If it happened every year, or every 3 years, or every 5th year, it would not be drought. Drought is an unpredictable condition that seriously impairs the viability of the food production system to meet the needs of the community. We have discussed in earlier chapters how communities adapted to this by storing grain or building up herds or migrating. In effect they knew the risk was there, and they had some broad idea of how bad it could be and how they might offset that risk. Overall they decided that the risk was acceptable and that they could cope with it. This was because drought was the exception, not the norm.

What we have seen over the last 25 years, particularly in Africa, has been the seeming persistence of drought conditions insofar as the crops and animals have failed to match the community's needs an alarming number of times during this period. Coping mechanisms begin to collapse rapidly if drought persists. For instance, communities will have to eat their seed corn or start to kill off their breeding animals, thus destroying their capital and their very basis for future well-being. Instead of areas recovering after the episodic drought, they only partially recover or are rendered useless and simply do not recover. Their productive capacity is lowered, sometimes below any level of usefulness to the people. They become less and less prepared for each succeeding onslaught.

In essence drought seems to become the norm, and consequently the risk of crop failure rises to unacceptable levels. The traditional coping strategies no longer provide sufficient insurance to give the community any confidence in its prospects for survival. Thus the people are left weakened and so less able to deal with the next drought when it arrives. The problem is not that droughts have begun to appear in these areas; it is that they have become unmanageable, persistent, and seemingly irreversible. But is this something new?

For the scientist, policymaker, and community at large the problem is whether this situation is part of a trend. A glance at the historical record shows that desertification has certainly occurred before. The Roman, Sumerian, Nabatean, and Babylonian civilizations all suffered from what we would now call desertification. Nigerian farmers often cultivate what are really "fossil" dunes, created at a time, around 20,000 years ago, when the Sahara was 300 miles south of its present limits. Former lake levels on all continents show evidence of wetter and drier periods. Thus we know that these changes can be profound and lasting, though we still have little idea about causation. We also have ample evidence that much of the Sahara was considerably wetter than it is now. We may see rock drawings at Tassili in southern Algeria that show people hunting animals long gone from that desolate area. We find the bones of extinct water-dwelling animals and even find the desiccated seeds of crops not grown or notable to be grown in the area for centuries.

Some Dilemmas

At this point it is possible to see some of the problems that are going to arise in interpreting this situation and then in doing something meaningful about it. First there is the question of whether desertification is something that happens from time to time as an inexorable part of the climatic, rather than the weather, process. Since we have evidence of fossil dunes and earlier desiccation, why is this not just a periodic reassertion of some natural cycle? If this is a cycle, how long will it last? Is it a cycle or a trend? Will recovery *ever* come? Is this a natural process that is localized in causation, or is it part of a global climatic change, possibly part of the "greenhouse" phenomenon? Then again, can the "evidence" of natural degradation, or desertification, result from human activities? Is it not the case that people can produce desert-like conditions without the intervention of any climatic change? Or can people, through their activities, produce the climatic changes that produce desertification? As though this were not byzantine enough, we may also have to face the question of whether this problem is getting worse or just getting better recorded. We know there were great and terrible droughts in the past (1910 to 1915 in West Africa, for instance). But now we have instant global communication and perhaps we are confusing being better informed with being part of a spiraling decline. It is a well-established tradition among grandparents, for instance, to tell you that summers were always better "back then", and that people had better manners, and so on. This tends to gloss over the two world wars, terrifying dentistry, painful childbirth, and a host of other things we are happy to be without. So in science, there is a danger when the past record is composed mainly of anecdote, oral history, and the like, and not hard data. This looks like a labyrinth, and what makes it more tortuous is the fact that a meaningful approach to desertification is unlikely to emerge if we do not clearly understand the *process* that is causing it. This brings the discussion to the question of desertification as a process.

Desertification as a Process

It would be helpful, right away, to say that desertification is not a *unique* process. Rather, it is the general condition of environmental degradation occurring in a particular place — the dryland areas. Rhodes, writing in *World Development* in 1991, states:

> The erosion of the concept of desertification, therefore, is counterproductive to the extent that broad use of the term may divert attention from inappropriate land use practices such as range mismanagement and overirrigation, which have been

the subject of intense scrutiny under different labels. . . . As Nelson, an environmental analyst for the World Bank, laments, "if it were not for the penetration of the word desertification in the literature we would have preferred to ignore it entirely and stick to the words dryland degradation."[1]

In those areas the effects of wind and water erosion, soil exhaustion, and so forth take the ecosystem below the baseline, producing a distinct end product called a desert. There is nothing to suggest that this is a process that is causally different from other processes. What distinguishes it is the end product. So, desertification is part of a broader issue of environmental degradation. It should, therefore, be seen in the context of the much broader struggle to conserve resources and sustain production everywhere.

In 1987 the United Nations held a global conference on desertification in Nairobi, Kenya. While this gave legitimacy and recognition to the problem, it also tended to give it a separate identity, and an institution was created to help deal with it. This focusing of attention on desertification as something unique tends to put it in a box alone and to disaggregate this one aspect of a much larger global issue of environmental decline that requires our *undivided* attention.

In looking for causation it is necessary to seek some element in the ecosystem equation that has changed. Bearing in mind the rapid advance of desertification — the United Nations have estimated that 5.5 million hectares of land are degrading toward desertification annually — we need to look for something that has changed *considerably* and *relatively recently*. This may still, of course, be a natural change of a cyclical form, since we know that some of the minor ice periods came on in as short a time as 70 years. But we do need to identify what it is that is causing the decline in productivity and the expansion of unproductive regions.

In terms of causation we are left with three broad categories of possible culprits:

1. Climatic change on a global scale.
2. Destruction of the vegetation cover by accelerated mismanagement caused by people.
3. An interaction between (1) and (2) so that the actions of people result in the *local* alteration of the micro- or mesoclimate. This is a phenomenon known as **biogeophysical feedback**.

Global Climatic Change

The general arguments pertaining to this phenomenon have been examined in Chapter 2, and they will not be restated here. If the world were subject to some major global atmospheric change, then boundary climates such as those along the dryland margins of cultivation and pastoralism would be particularly susceptible to change. Such evidence as we have, and it is quoted in Chapter

2, would seem to suggest that drier areas in West Africa and Asia would, if anything, get moister with global warming. A general weakening of the vigor of the atmospheric circulation would tend to lead to an equatorward expansion of pressure systems. This would, in the case of the great deserts of Africa and the Middle East, lead to a southward expansion, in line with what some scientists think they see.[2] But we have no evidence to support a thesis of a general decline in system vigor. There is an additional problem. We measure desertification as a process against some "baseline" or "norm" — in this case the condition of the drylands when scientists, colonists, and others started taking serious recorded note of the conditions. As late as the 1960s, the "30-year norm" was still widely taught. It stated that if you had 30 years of weather records over a broad area, you had typified it. This rationale implied a level of stability that we would not hold to today. On this basis, some scientists now argue that the records, taken as baseline in West Africa and parts of Asia before the last war, were taken during an abnormally humid phase. Indeed, it may be that agricultural cultivation expanded during that more humid phase into areas that are "normally" inimical to such activities. What we are seeing now would, under this hypothesis, be a readjustment away from an exceptional situation, not the beginning of one. Last of all, how long does a drought have to go on before it ceases to be a prolonged or extended drought and becomes a climatic change? Without historical records extending over hundreds of years, this is a hard question to answer. So, in this context, desertification becomes a readjustment toward something more "normal" and "stable". It is clearly a horribly difficult problem to resolve.

Although, as we have said, we cannot prove that climatic change causes desertification, what would be the implication of such a causal relationship being proved? First, the cure may well be outside the hands of the tropics since it may involve industrial pollutants generated by the developed, temperate areas. Second, policymakers tend to be overwhelmed by, or choose to be overwhelmed by, "acts of nature", sometimes because these absolve the politicians of any blame. Once something becomes a natural disaster, the politician is provided with an easy exit. As of now we have no conclusive model to explain desertification in global climatic terms, which is not the same thing as saying "that is not the cause." However, scientists are reluctant to commit themselves in such an uncertain and unproven environment to giving usable advice.

Mismanagement

This takes us back to the broad ideas surrounding **overpopulation** discussed in Chapter 7. In the case of the drylands, overpopulation may occur at astonishingly low densities such as one or two people per square mile. These ecosystems are extremely fragile, which is another of their "marginal" qualities. Stress accumulates quickly. Overpopulation may result from any of the following: a growth of human population due to better medicine, the cessation of warfare, or the closure of migration routes across new political boundaries; the buildup of herds and flocks to support more people, and as a result of better groundwater provision and veterinary medicine; greater demand for land to meet needs beyond subsistence (taxes, etc.); the need for more and more land to feed a human or animal population in a degrading environment (a spiral of decline); the demands for wood as fuel; and the expansive pressures of neighboring cultivators.

If one or any combination of these factors starts to extract more energy from this ecosystem than it can locally replace, then, like any ecosystem, it will start to degrade. Many of the physical manifestations of this decline will be remarkably similar, and sometimes identical, to the phenomena observable as a result of additional aridity. Increased runoff due to compaction and surface-stripping will, for instance, prevent groundwater recharge, and so wells will go dry, oases may wither and decline, the range of cultivable crops may diminish, and the trees whose roots now lie above the available groundwater may perish. It is quite reasonable to argue that the tremendous expansion of cotton cultivation under French pressure in West Africa during colonial times left the land in countries such as Mali and Niger bare for critical times of the year and also prone to disastrous outwash when the rains came. Furthermore, continual monocropping weakened the structure and resilience of the soils making them more *vulnerable* to erosion in their dried-out and bare condition. In Saudi Arabia the enormous demands of cities such as Jeddah drew down the groundwater in its Wadi Fatima hinterland to such an extent that ancient oases dried up and died, resulting in an aura of total desiccation. At the same time, in neither place was there any serious evidence of climatic change, even as the landscape became visibly and dramatically more desert-like. At least, if the problem arises from elements of mismanagement through overuse, then there is the possibility of locally generated solutions and readjustments. These, however, may involve difficult choices for the local political establishment — such as we saw in the case of the marginal lands in Kenya — and in that context an "act of God" becomes an attractive political alternative.

Biogeophysical Feedback

There are several ways, *theoretically*, that people on the ground may influence the climate above them — or the climate passing over them in the case of the monsoonal climates of the drylands. It should be stressed that some of these cases lead mathematically to local climate change in elaborate computer models of the atmosphere, but that is no proof of causation. Some elements of biogeophysical feedback were outlined in Chapter 2, but two are worthy of mention here. To take the Mali example again, if cotton cultivation expanded over a great area it would, before the crop growth was established each year, leave bare ground exposed. This then increases the **albedo**, or reflective index, that bounces heat back into the atmosphere and reduces the heating of the ground surface. This would in turn reduce the capacity of the ground surface to trigger the convection effect on monsoonal air passing over it and so the moisture would not be released.

Similarly, exposing the cotton fields to the desiccating effects of winds results in a fine tilth being swept up into the atmosphere. This may then form a dust-laden layer that absorbs the incoming radiation and heats up much faster than the surrounding air above and below. Convected, moist air rising as a result of surface-level triggering would then run into this "hot" layer and cease to rise. In both these cases the rainfall is constrained, the monsoon does not arrive, crops fail, and the area starts to look like a desert.

Nomadic Pastoralists: A Sad Case of Cause and Effect

A particular case will illustrate how the various elements of cause and effect become scrambled, confused, and sometimes totally reversed in the interpretation of the destruction of grazing lands.

Large parts of East Africa are too dry for cultivation and so have become home, over the centuries, to numerous nomadic tribes such as the Karimojong of Uganda or the Maasai of Kenya and Tanzania. Many of these grazing areas have suffered serious damage during this century, though they have not been ravaged to the same extent as their counterparts in West Africa.

To comprehend what has been happening in East Africa it is necessary to reiterate a few points made in Chapter 6. Pastoralists raising animals for *subsistence* in unimproved dryland conditions rely on them for what they can produce as food on a regular basis (milk, blood), not as meat producers except under extraordinary circumstances such as for weddings and funerals. At any given time only some of the animals will be producing milk, and some of that will have to be shared among the calves. Furthermore, because of poor grazing conditions, the amount of milk produced per beast is going to be meager indeed.

In addition, the herders have to cope with resources that are geographically dispersed and episodic. Animals need both grazing and water resources that are relatively contiguous since cattle have a low tolerance for going without water for any length of time. In order to cope with this problem, which changes dramatically season by season, the herders must have an encyclopedic knowledge of the local environment and geography.

Over and above the day-to-day and seasonal adjustments and risks, there is the constant threat of drought. It is to this threat that the coping strategy must ultimately adjust. And so the herders will attempt, quite naturally, to keep as many *productive* or *potentially productive* animals as they can. This provides the herder and the herder's family with the best *insurance* possible under that technology. It is this need for survival that gives rise to what has sometimes been called "the numbers mentality" of the pastoralists. There is no doubt that they do have a numbers mentality, but it is selective and totally rational.

It happened in these pastoral areas, as in most other parts of the tropics, that Western technology and administration came along with dramatic suddenness,

1. The "Traditional" Human Ecology

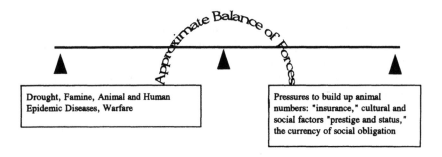

2. The Intervention of Western Technological Change

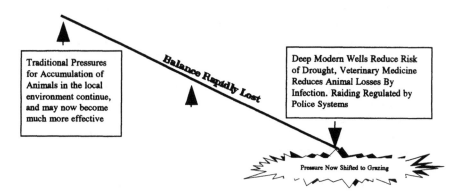

Figure 15.3. The impact of "development" on natural checks and balances in pastoral societies.

and the consequences of the arrival are shown in stage 2 of Figure 15.3. From the perspective of the administrators who were eager to commercialize the indigenous economies and provision the growing administrative centers with meat, it seemed that these pastoral people had two distinguishing characteristics: they had more animals than they needed (which might be possible in a good year), and they suffered heavy losses when the droughts came along. There appeared to be an obvious answer here. Someone once said that "for every major problem there is a solution that is obvious, easy, and absolutely wrong." In this case logic seemed to suggest that if new technology was used to tap deeper groundwater resources than the indigenous communities could reach with their traditional methods, then fixed, permanent water points could be constructed. These could be placed close to unexploited grasslands where previously a lack of surface water had constrained grazing. This would then expand the resource base available to the pastoralists. Also, by eliminating the killing effects of drought, it would be possible to increase the "surplus" of animals available for offloading into the monetized market system. And so huge groundwater development programs began dotting the landscape of the semiarid areas.

There were, however, several flaws with this argument, and most of them derived from the imported nature of the interpretation that had been made of the pastoral system. In the first place, if a change enables more animals to survive in a system characterized by low productivity and high risk, then it is likely that the people will retain the increase within the family's control to offset the risk. This would be the typical response of the coping mechanism (see Figure 15.1). Second, the people would be likely to sell stock only insofar as the perceived benefit of the sale would offset the perceived loss of productive capacity and subsistence. In these remote areas opportunities to buy anything are fairly limited. Furthermore, selling productive animals would mean putting that part of production into the hands of merchants and suppliers and into the money system. All of these elements are unreliable, and risk avoidance is then made dependent upon a third party. Furthermore, over the centuries, the livestock have come to play a variety of social and economic roles that money may well not replace overnight. The real question is, given this opportunity to increase herd size, will the pastoralist then respond to pressures or opportunities to offload the increase on the market? The answer has generally proved to be no.

At the same time as water development was proceeding, a further technical change occurred that influenced the ecological equation. One reason that herders were so conscious of livestock numbers was the fact that the herds and flocks tended to be ravaged by epizootics such as east coast fever, trypanosomiasis, and rinderpest (though the last was an unexpected side effect of colonialism). Modern veterinary methods permitted the wholesale vaccination, or dipping, of animals against these diseases. Consequently the second major constraint on animal numbers was diminished, and the herds were poised on the edge of a population explosion.

As numbers grew the pressures on grazing also grew, and the colonial administrations became increasingly frustrated with the pastoralists for not selling their animals and behaving "rationally". The increasing numbers of animals, and the numbers of people that the animals had to support, led to greater and greater pressures on grazing resources, especially in great bare circles around the new water points. Eventually the main stress point shifted from an absence of available water during droughts to an absence of available grazing as the grasses were selectively grazed out, and as unpalatable species colonized the overgrazed areas. As grazing diminished, the risk of loss to the increased herds grew. The traditional reaction to this was to *increase* the pressure to have more animals (more "insurance") and to reduce the number of animals being released into the market. So the crisis fed upon itself in a reinforcing manner.

The degradation was steady and relentless. What was happening, however, was that the grazing areas were diminishing in their resilience or their capacity to withstand pressure. As always, it was the extreme events that showed up the crisis. Therefore, when droughts came, even if they were climatically no more severe than previous visitations, their effect was significantly worse. There were more people to be supported on the diminishing resource base, and that resource base did not have the capacity to withstand the drought as well as it did in the past. Plants died much faster, ground was laid bare more quickly, and available grazing areas diminished dramatically. Drought that would have been within the purview of coping mechanisms previously now led to famine: man-made famine. It is this famine which caused the widespread loss of life during 1968 to 1972 and 1982 to 1984, not more severe droughts.

The authorities accused the pastoralists of being "conservative", resistant to change, and even of being the agents of their own destruction, which is the ultimate form of irrationality. Meanwhile the grazing areas declined dramatically. To have thrown their risk on the market, instead of retaining it as they did, the pastoralists would have had to resist strong cultural constraints and change their diet since fresh milk is unlikely to be available in these areas from any other source. These communities were not actually resistant to change as such, they were simply facing systemic change at a pace and scale for which nothing in their traditional lifestyle had equipped them. It is hard for a community to give up perceived control over its own destiny, and that is what sudden commercialization meant in these remote places. It meant now that either the government famine relief organization or the merchant in his shop were the foundations of the community's ultimate survival. Neither of these was trusted, often with good reason.

Unfortunately the problem was now bedeviled by a large perception gap. The often urban-based or expatriate administrators saw the pastoralists as stubborn and reactionary. The pastoralists, in turn, saw government marketing schemes, taxes, etc. as a form of robbery and deprivation of control over the basic, traditional means of survival. Under pressure, the pastoral communities resorted to time-honored practices of raiding or illegally migrating across the boundaries of other communities, or even countries. This made the administrators even more resolute in their aims to "settle" the pastoralists and turn them into stable, governable, controllable communities for "their own good".

Alas we have progressed little from this point. The extended droughts from 1968 until the present time over much of the drylands of the world have highlighted the parlous state of the communities living there. Many have already migrated to equally hopeless urban situations. However, the situation is still often perceived as an ongoing drought or "natural crisis". It would seem to have passed the point where we can maintain that exceptionalist stance, and responsibility seems to lie more with people than with nature.

Conclusion

Even though desertification seems to be a labyrinth of unprovable possibilities, the general thrust of global climatic evidence would seem to suggest that whatever *is* going on with the general atmospheric circulation, it is probably *not* the cause of desertification. The answer lies in the buildup of pressure on the most fragile of environments: too many people and too many animals on a degrading life-support system. At the same time, the extremely low level of resources, the lack of alternative activities, and the distrust between government and communities in these areas make solutions difficult to find and implement. Indeed, following the crippling drought of the 1960s and 1970s in West Africa, *every* government in the affected region collapsed, making an effective response even more difficult. Once more we need to seek a sustainable solution, a difficult task since we cannot use averaging strategies in systems characterized by periodic high risk. What we do need to do, however, is pay more attention to the traditional coping mechanisms, and at least be sure that everyone is convinced that the level of risk avoidance of any new strategy of land management is at least as good at the strategy it replaces. At the same time, there will be cultural barriers to change because societies find security and reassurance in the unique way that they do, and have done, things. Sometimes it is difficult to distinguish cause and effect in culture, too: Do people have large herds because they "love cattle"? Or, do they love cattle because large herds are traditionally a sign of security, well-being, and an ability to avoid hardship?

The desertification situation demonstrates all too well the paradox of two communities conducting an empty dialogue, where neither really understands the basis for the other's actions. The victim of all this tends to be nature. To control desertification, it will be essential to maintain balanced ecological practices that cope with periodic peaks of risk, and these are not easy to construct. Nevertheless the Israelis, for example, have been successful in reviving the water-harvesting methods of the long-gone Nabatean population. But it is easy to see how policymakers find it easier simply to avoid this problem.

Once more the technology to combat desertification exists. It is not as though we have no idea how the situation may be arrested and reversed. The great gap is in the available financial resources, and the will to commit the financial resources necessary to implement these changes as manifested by (1) the commitment by world governments to fund the activities of the United Nations in combating desertification, (2) a lack of funding to cover the recurrent or ongoing struggle (it is fine to plant trees, but they have to be nurtured in their early years or they die), (3) the resolution to formulate and enact policy for some of the poorest people in the world, and (4) recognition that famine is not an act of God but is an act of people — often people who have no alternative but to bring about their own destruction in the present circumstances. Disaster, it seems, is now a way of life in these marginal areas. Only a coordinated *world* effort will resolve this issue of the otherwise irretrievable loss of productive resources.

REFERENCES

1. Rhodes, S. L. "Rethinking Desertification: What Do We Know and What Have We Learned?" *World Dev.* 19(9):1137-1143 (1991).
2. Crosson, P. R. and N. J. Rosenberg. "Strategies for Agriculture," in *Managing Planet Earth* (New York: W. H. Freeman, 1990).

OTHER USEFUL READING

- Baker, R. "The Sahel: An Information Crisis," *Disasters* 1(1):13-21 (1977).
- Eckholm, E. and L. R. Brown. "Spreading Deserts: The Hand of Man," Worldwatch Paper #13 (Washington, D.C.: Worldwatch Institute, 1977).
- Glantz, M. "Drought in Africa," *Sci. Am.* 256(6):34 (1987).
- Grainger, A. *Desertification: How People Make Deserts, How People Can Stop, and Why They Don't* (Washington, D.C.: Earthscan, 1982).
- Heathcote, L. *The Arid Lands: Their Use and Abuse* (London: Longman, 1983).

CHAPTER 16

Prospects for a Sustainable Future in the Tropics

There is nothing new or different in what you have read so far. What is unusual is that the various components of this study do not usually appear *together* between the covers of one volume: the indisputable interrelations and boundaries of natural science as far as we know them, the more pliable matter of social sciences, and the altogether malleable substance of history are rarely caught in the same neighborhood, never mind cohabiting in rare intimacy as they do here. But that is how life is, and hard though it may be to try to reconcile scientific proof with opinion and the political process of consensus building, this is what we must do if we are to survive, and a great part of the human population is having a hard time doing that right now.

The incorporation of the historical perspective suggests one overriding value above any others it may yield. It teaches us that the division of the world into rich and poor regions — specifically into the white temperate and black tropical worlds — is something that was induced. It is not, and was not, the "natural order" of things. A *process* occurred that beggared some of the greatest cultures and economic systems of antiquity, and set in train the very uneven allocation of wealth and technology that now typifies our global "order". Through slavery, the colonial system, and asset stripping, the West (including the Russian and Soviet empires) grew mightily rich, distorting and stunting the progress of almost all tropical dependencies in the process. What is important is recognizing that this occurred, *not* apportioning guilt or blame, or trying to construct a new world order based on troubled consciences. We need something much more positive than guilt to frame our actions. The value of understanding this historical process is to recognize the damage done by dividing the world and by exploiting it for the short-term gain of the few, both of which still occur. Since the division was created by an historical process, it will take an historical process to change it.

Thus, the West should dispense forever with the idea that the tropics are poor because they are intrinsically poor. They were made poor. We must be prepared to work toward their full and fair reincorporation into a world order. Proud words, but why should anything change now?

Samuel Johnson, who said so many wise things, remarked that the prospect of being hanged "focuses the mind wonderfully." It seems that humankind needs some form of shock or crisis to make the big decisions. Where we are now is at the edge of a big decision — what one writer once referred to as "a hinge of history" around which all events turn and direction changes irrevocably. What is this hinge? Why now? And what does this mean for the tropics?

What is meant by a hinge of history is a radical change or paradigm shift, during which we do not just play around with the relative value or composition of the variables, but we change the way we think about thinking — what might be called "a new world vision". Many thought Marx had produced such a vision, and intellectually he had. However, it did not sit well with the essentially free spirit of human individualism, and replaced the individual familial, or community force for change with, of all things, a monumental bureaucracy. Marxism also ravaged the earth and left damage as yet unmeasured.

The answer to the question "why now?" comes from the logical progression of the development of science and technology. The harnessing of energy into ever-more powerful machinery has enabled humankind to bring about change at an accelerating pace to the point where we can now turn ourselves into vapor at the push of a button if we so wish or render huge areas uninhabitable as around Chernobyl. This increasing power has led to a similarly accelerating ability to transform the tropical environments through logging, civil engineering, fertilizing, and so on. The West can now effect damage at a scale and speed previously unimagined. Progress in medical science and technology insulate more and more of us, even in the poor tropics, from the natural checks and balances on exploding human population numbers. In short, we have the capacity to bring about both *regional* and *global* change in the physical environment — below us, around us, and above us. For decades the developing countries of the tropics have argued that they had no countervailing power against the wealth and might of the industrialized nations. Once, briefly, there was an illusion of a shift after OPEC and the Group of 77 used the oil card following the Yom Kippur War of 1973 to 1974 — but that did not last. Now, however, the power to effect change may finally have shifted.

At this point in history, the burning and clearing of the Amazonian rain forest, as well as the emission of CO_2 from industrialized countries, threatens the sustained utility of the global atmosphere for all humankind. The ability of *all* of us to damage everything for everyone might just be what is needed to give substance and reality to what George Bush would, no doubt, call the "shared vision thing". The consequences of global warming, ozone depletion, rising sea level, acid rain, etc. are not influenced by those invisible lines on the ground that have bounded our nations and governed our actions for so long. Inside these lines until now we have created our sovereign laws and made our policies. These lines defined our world.

The sanctity of sovereignty is now under siege, and its demise will be a major advance for the long-suffering globe in general and the tropics in particular. First, it is hard to see how we may rise to the challenge of meeting global environmental crises using the fragile construct called international law. International law relies on the accused to *accept* the standing of the court and the judgments it hands down. Imagine how long sovereign national law would last given the same options. Once things looked as though they were not going your way, you would simply declare the court incompetent and leave. That, in effect, is the reality of how *nations* behave under the existing conditions of international law. So there are no effective sanctions because the nations of the world have never recognized the sovereignty of anything greater than the nation itself — look at the history of alliances, of the League of Nations, for example. Now we are seeing *radical* changes in sovereignty. On the one hand the emergence of the European Community is a fundamental transfer of sovereignty, the United Kingdom notwithstanding; on the other hand the actions of the United Nations in Iraq following that country's invasion of Kuwait would have been unimaginable just a few years ago. The Iraqi protests of "interference in sovereign affairs", which would have sent the UN scampering formerly, were just, literally and figuratively, cries in the desert.

Second, the end of the Cold War creates the possibility of a non-bipolarized view of the future, allowing the "common" referred to in the 1987 report by Norwegian Prime Minister Gro Harlem Brundtland, *Our Common Future*, to have some real and practical substance. Not only does the new era allow for attention to a shared purpose, but the reduction of the military threat allows for the transfer of financial resources and that well-known military organizing ability toward saving the planet rather than blowing it to pieces. So the industrialized world would therefore seem to have the perception and the means for global action.

Third, we have restored confidence in the UN to perform the role of steward of shared sovereignty in some matters of common concern. This is not because the UN has become any "better" or indeed any different. What has changed is its former preoccupation with ideological mind games that frustrated and emasculated the organization for the 40 years following the Second World War. This end to posturing was coupled with a realization that we are all in the same boat — one that appears to be sinking. The UN now serves as a real forum for global action. Perhaps one of the most interesting aspects of this change is the proposal being prepared by a coalition of environmental groups for the 1992 UN conference in Rio de Janeiro on the global environment. This document, in the form of a petition with millions of signatures, starts with the words "We the peoples . . . ," and is intended as an environmental charter. What is interesting is that, like the Constitution of the United States, it refers to people, not to kings or governments. This is the ultimate expression of the shared vision, the realization, that here, in the global environment, regardless of culture, geography, or political system, is something that unites us all. The UN Declaration on Human Rights, of course, sets out to do the same thing, but was, until the extraordinary events of the late 1980s, frustrated by the self-serving myopia of archaic nationalism. That world, it seems, is truly coming to an end despite the unfulfilled resolution of fossilized imperial structures such as Yugoslavia and the resolution of the great vacuum of the Soviet empire. This too shall pass.

All the preceding paragraphs seem to have taken us a long way from managing the tropical environment, which is supposed to be what this book is about. But not so. The message of this book is that the tropical environment has been the victim of unequal global incorporation ever since Christopher Columbus got lost on his way to the Indies and since the Portuguese "discovered" much of Africa, to the astonishment of those who had been living there for millennia. The tropics now has to cope with the unreconciled clash of European and indigenous systems, as well as the legacy of a grossly unequal incorporation of one part of the world into an economic system created by and for the other half. Here are the keys to real sustainability: access to resources and markets, fair recompense for labor, and a stake in a lasting future.

The buzzword in environmental circles for the moment — and let us hope it will be the last because it is the only one that matters — is *sustainability*. The meaning is simple: "produce in such a way that meeting your needs does not imperil the capacity for others to meet their needs later," or "don't spend all the capital on a wild party." That, in effect, is what the West and the East have been doing for over a century. Unfortunately only a small percentage of the world was invited, but everyone had to pay for a ticket.

What are the lessons? Having presented a broad, interdisciplinary, historical view of environmental management in the tropics, we turn to the priorities for future action.

Economics and Ecology

As the former head of the World Bank, Barber Conable, observed, "Sound ecology is sound economics." As they say in management circles, the priority must be "to operationalize that concept." We have observed how conventional economics does not incorporate any value for resources lost; it merely adds whatever is done with these resources to the stock of value. Project appraisal and cost/benefit analysis *must* incorporate the cost of these losses — a difficult task with something as unknown and *potentially* valuable as the tropical rain forest. Work is being done in many industrialized countries and multilateral organizations to internalize the environmental components of conservation, sustainability, and the cost of resource loss into conventional appraisal methodologies. Economics is, after all, just a tool to be used in calculating options within the framework of human values. Those values, up to now, simply ignored the realities of time and natural science. Formulating sustainable economics will be the true expression of changed reality. For 60 years the old system of national accounts has fed us indices of "progress" that were hedonistic, mortgaged the future, ruined our heritage, and were supposedly "rational".

International Economic Relations

We live in a divided world, and that division must be the target of our attention. The capital needed to close the gap is flowing in net terms, the wrong way — namely out of the tropics. This capital outflow, together with the albatross of debt, is not going to allow tropical developing countries to do anything other than take a short-term, survivalist view, for which nature will go on paying the price. Having been frozen out of much of the industrial revolution by the colonial and neocolonial systems, the developing countries have exported most of the added value of their raw materials to the industrialized nations. Thus, for example, only around 10¢/lb of coffee, or around 1.5 to 2.5% of the final selling price, is retained in the exporting country. The same picture holds true for timber and most other primary products. The real wealth accumulates downstream. Until the tropical countries can *truly* participate in the value-adding process, which means being free to export processed and manufactured goods without giving away the store through duty-free export-processing zones, they will remain the poor relations. From the perspective of the tropics, it is alarming to see the reconfiguration of the world into (protectionist?) trading blocs of "rich peoples' clubs" such as the European Community or the North American Free Trade Area. The inclusion of Mexico in the latter, however, would be a turning point, linking a developing country directly with the richest country on earth.

If there is to be real change in the management of the tropical environment, then several things are going to have to happen a long way from the tropics. To enable real capital formation and the creation of meaningful opportunities away from the land, serious industrialization and a fair deal with the developed countries will be required. This means global expansion of the free trade concept, a considerable amount of debt elimination, and capital availability on preferential terms — in other words reversing the present flow, opening the trade doors, and increasing concessionary lending. Without this there will be no way out of the short-term debt, balance-of-payments crisis, and nature will go on paying the bill.

Back home in the tropics, a similar liberalization must take place. International access to markets must be matched domestically by access to resources, which means making some painful political decisions about land tenure and redistribution. If tropical countries are asking for a "fair deal" from the industrial nations, these same countries must be prepared to give a fair deal to their own people and reverse the price twists that keep farmers poor and encourage the rape of the land. When developing countries cry out for more credit on easier terms they should also look at their own credit systems that effectively lock out their poorest citizens, who cannot secure a loan since they are considered to have no assets. But these are the very people who need loans the most — even $30 can make a major difference — and who have the best repayment record. And when tropical countries cry out for justice, they too must replace arbitrary and capricious government by representative and participatory systems. Thus, giving people a stake in the future, in countries that themselves are trying to secure a stake in the future, obviates the security imperative that creates the *need* for large families. This is not to say that the population issue can be dealt with at a stroke, but it does put population back into perspective as a *dependent* variable, where it belongs. The environment of poverty must cease to have dominion.

Changes in the Developed Countries

The environmental initiatives in the developed capitalist countries from the 1950s onward were a reaction to the unregulated creation of great wealth. Thus came into being what might be called the first generation of postwar environmentalism. This consisted of making the polluter pay, setting standards, and policing them, thereby internalizing environmental costs. All this was within the sovereignty of nations, and was, so to speak, the easy part. When it comes to global environmental management, which is the only realistic construct within which sustainable environmental management of the tropics is remotely likely, there will also be costs, and most of these will fall on the developed world, reversing a 500-year trend.

The resources needed to break the tropics out of the poverty trap and to provide a realistic alternative to the rape of the earth are stupendous. If humankind really does not want India to use greenhouse chlorofluorocarbons (CFCs) or China to burn its high-sulfur coal, then the industrialized countries are going to have to underwrite an alternative. Telling the industrializing countries to stop will not work. Similarly, the short-term earnings for East Asian governments from tropical hardwood sales have to be replaced. As with ivory and furs, the rich of the developed world must also overcome their craving for teak. It has been suggested that a tax on carbon, which is a by-product or component of most transformational processes, would fairly extract from those who use most resources ($25 billion per annum from OECD countries alone). This "carbon tax" is another version of the internalizing of environmental costs, or applying the "polluter pays" principle, but in a transnational context this time. There will also have to be a sharing of intellectual property if, for instance, an alternative to CFCs is to become feasible in India. At the moment, nine tenths of research and development is a Western occupation and possession. Those who invent and develop environmentally friendly products will expect to be compensated, naturally, but this barrier may well put the product beyond the means of its potential users in poor nations. In this context of selective access, a global environmental technology fund would seem, in one form or another, inevitable.

Possibly the greatest contribution the rich countries could make in the short term is to attack the waste that accounts for a scandalous 30% of world resource consumption at present. Then some fundamental questions about consumption patterns and a rejection of the maniacal pursuit of postwar consumerism can follow and more thought can be given to equity, humanitarianism, and sustainability: the real "quality of life". Look around and see how many pens or clothes you possess, for instance. This is madness. You could survive with fewer. Try it. It is an astonishing fact for us to realize now, but as late as 1945 the average Englishman owned *one* pair of trousers! Taxes on ozone depletion, carbon use, or even energy will all help to raise consciousness and provide resources for real change across the globe, perhaps to fund research and technology transfer.

Research

We have already remarked on the need for greater and freer access to the products of ecologically sound research by developing countries. There are other imperatives for research, however. It is necessary that we make every effort to record and understand ethnoscience and to make up for the terrible cultural paternalism, and worse, that led the West to regard the tropical societies, economies, and cultures with contemptuous indifference. Not only in terms of plants and fauna is there much waiting to be discovered, but also in terms of native medicines, indigenous taxonomies, oral histories, and religious world views. Even in their traumatized and bulldozed conditions, the indigenous societies demand our respect and sincere curiosity as partners.

Some time back it was fashionable to speak and write about "appropriate technology", which was to say some alternative to the wholesale appropriation of late 20th-century Western technology from the temperate, capital-intensive realm. This appropriate technology was related to and generated for the realities and needs of farming in specific economic and environmental conditions. Such remarkable innovations as the minimum tillage tool bar, adapted for animal-draft use in Botswana and the Sudan, could often produce startling changes in the productivity of peasant labor, affording it one real opportunity to break out of poverty. But these inventions are almost never seen. Why? Because the local authorities who had to approve and promote these technologies considered them "old fashioned", "second-class", and an attempt by the West to keep the tropical countries backward. They were no such thing, and the subsequent alternative, the wholesale purchase of tractors, was economically disastrous. But this sadly is the psychology of catching up — not with yesterday, but with someone else's tomorrow. The priority must be to generate practical and indeed appropriate solutions to the needs of the great mass of people on the land where they farm. This means a continual refocusing of research on the basic production system; looking at social and economic realities at least as much as the soil, plants, and water; and — most of all — producing change that the local environment *and* the local society can sustain. Rarely have either of these conditions been realistically addressed or met. The tropical countries must recover a genuine psychology of self-respect, not an aggressive reactionary posture.

The Energy Pathway

In the last few decades, change in the tropics was largely accompanied by energy mining — the exhaustion of soils in cleared forest areas, the burning or export of trees, the erosion of the organic top horizon of many tropical soils, the wholesale stripping of woodland for fuel, and so forth. New approaches to tropical agriculture and land management *must* focus on sustainable energy pathways. This means providing, for instance, managed woodlots to keep families provided with fuel without their having to denude all the land in sight of trees. It means providing a substitute for the burning of animal dung, which is needed in the nutrient cycle. It means growing successions and interplanting crops that maintain the fertility of the soil. It means applying genetic engineering to non-capital-intensive solutions for improving food crops, and it means shifting tropical agriculture away from an increasing reliance on nonrenewable hydrocarbons.

And Finally . . .

The real challenge we all face is to accept the responsibilities of the new order *now*. We are seeing great change — extraordinary reconfigurations of states, the end of bipolarism — but at the same time the growing division of rich and poor among and within nations. However, the spread of democracy and its empowering reach, the global village of instant communication, and even the rise of English as a world language offer real hope for common purpose. As the millennium approaches, maybe we should think in millennial terms, casting aside much of the baggage of nationalism, sovereignty, and greed. The tropics faces the millennium in bad shape; its people and its natural resources are in serious trouble. As we recall the quincentenary of Columbus's voyage, we should take stock of the historical legacy following his arrival and mark the occasion by closing the book on a sad, exploitive episode in terms of both intercultural relationships and humankind's relationship with nature. The sheer scale of the challenge should offset the other realities of recession in the industrialized nations. Countries have shown a willingness, as illustrated by the signing of the Montreal Protocol of 1987 limiting the production of CFCs, to place *potential* risk over *proven* damage. This is a remarkable example of politicians thinking beyond tomorrow. Perhaps we can take heart from that. The environment management of the tropics will remain a chimera until the global perspective of stewardship becomes a practical reality. But time is very short.

USEFUL READING

- Brown, L., et al. *State of the World*. (New York: W. W. Norton & Company, 1991).
- Caldwell, L. K. "Globalizing Environmentalism: Threshold of a New Phase in International Relations," *Soc. Nat. Resour.* 4:259–72 (1991).
- Conable, B. B. Address to the World Resources Institute, World Bank, Washington, D.C., 1987.
- Durning, A. B. "Action at the Grassroots: Fighting Poverty and Environmental Decline," Worldwatch Paper #88, Washington, D.C.: Worldwatch Institute, 1989.
- Eckholm, E. "The Dispossessed of the Earth: Land Reform and Sustainable Development," Worldwatch Paper #30, Washington, D.C.: Worldwatch Institute, 1979).
- Ewel, J. J. "Designing Agricultural Ecosystems for the Humid Tropics," *Annu. Rev. Ecol. Sys.* 17:245-271 (1986).
- International Monetary Fund. "Balancing Development and the Environment," *Finance Dev.* 26:4 (1989).
- McKibben, B. *The End of Nature* (New York: Anchor Books, 1989).
- MacNeill, J., P. Winsemius, and T. Yakushiji, Eds. *Beyond Interdependence: The Meshing of the World's Economy and the Earth's Ecology* (New York: Oxford University Press, 1991).
- Pearce, D., A. Markandya, and E. B. Barber. *Blueprint for a Green Economy* (London: Earthscan Publications, 1989).
- Pluckett, D. and N. Smith. "Sustaining Agricultural Yields," *Bioscience* 36(40):5 (1986).
- Riddel, R. *Ecodevelopment: An Alternative to Growth Imperative Models* (Hampshire, England: Gower, 1981).
- Schramm, G. and J. J. Warford, Eds. *Environmental Management and Economic Development*. (Baltimore, MD: Johns Hopkins/World Bank, 1989).
- Seager, J., Ed. *The State of the Earth Atlas* (New York: Touchstone/Simon and Schuster, 1990).
- Stark, L. *Signs of Hope: Working Toward Our Common Future* (New York: Oxford University Press, 1990).
- Thibodeau, F. R. and H. H. Field, Eds. *Sustaining Tomorrow* (Hanover, PA: University of New England Press, 1985).
- Tolba, M. K. *Sustainable Development* (Guildford, England: Butterworth, 1987).
- World Bank Development Committee. *Environment and Development: Implementing the World Bank's New Policies* (Washington, D.C.: World Bank Development Committee, 1988).

Index

Stockholm conference, 117
United States
  agriculture, 50, 51–53
  beef consumption, 175
USSR, agriculture, 115–116

V

Vegetation
  climax, 1–2
    classification, 28–29, 30
  desertification, 184
  growth and temperature, 25–26, 33
  indicator, 49
  major types, 3
  rain forest, 169
Victorian era, 105
Volcanic soils, 40

W

Water harvesting, 40
West
  attitudes toward rain forest, 168

environmental conservation, 123
West Africa, see also Sahel, specific
      countries
  climatic change, 32
  desertification, 184
  intertropical convergence zone, 23
Western agriculture, 43
Western aid characteristics, 117–119
Western science
  colonialism, 104, 106
  research, 129–131
West Indies, colonialism, 96–98
Wheat production, 130
Wind, 26, 28, 194
World Bank environmental policy, 151
World Resources Institute (WRI), 122

X

Xerophytic vegetation, 28, 184

Z

Zambia, 58

Milton Keynes UK
Ingram Content Group UK Ltd.
UKHW040101071024
449327UK00019B/713